T0227986

Designing Ergonomic, Safe, and Attractive Mining Workplaces

Designing Ergonomic, Safe, and Attractive Mining Workplaces

Joel Lööw
Bo Johansson
Eira Andersson
Jan Johansson

CRC Press
Taylor & Francis Group
Boca Raton London New York

CRC Press is an imprint of the
Taylor & Francis Group, an **informa** business

CRC Press
Taylor & Francis Group
6000 Broken Sound Parkway NW, Suite 300
Boca Raton, FL 33487-2742

© 2019 by Taylor & Francis Group, LLC
CRC Press is an imprint of Taylor & Francis Group, an Informa business

No claim to original U.S. Government works

Printed on acid-free paper

International Standard Book Number-13: 978-1-138-09221-1 (Paperback)
International Standard Book Number-13: 978-1-138-09273-0 (Hardback)

This book contains information obtained from authentic and highly regarded sources. Reasonable efforts have been made to publish reliable data and information, but the author and publisher cannot assume responsibility for the validity of all materials or the consequences of their use. The authors and publishers have attempted to trace the copyright holders of all material reproduced in this publication and apologize to copyright holders if permission to publish in this form has not been obtained. If any copyright material has not been acknowledged, please write and let us know so we may rectify in any future reprint.

Except as permitted under U.S. Copyright Law, no part of this book may be reprinted, reproduced, transmitted, or utilized in any form by any electronic, mechanical, or other means, now known or hereafter invented, including photocopying, microfilming, and recording, or in any information storage or retrieval system, without written permission from the publishers.

For permission to photocopy or use material electronically from this work, please access www.copyright.com (http://www.copyright.com/) or contact the Copyright Clearance Center, Inc. (CCC), 222 Rosewood Drive, Danvers, MA 01923, 978-750-8400. CCC is a not-for-profit organization that provides licenses and registration for a variety of users. For organizations that have been granted a photocopy license by the CCC, a separate system of payment has been arranged.

Trademark Notice: Product or corporate names may be trademarks or registered trademarks, and are used only for identification and explanation without intent to infringe.

Library of Congress Cataloging-in-Publication Data

Names: Loow, Joel, author.
Title: Designing ergonomic, safe, and attractive mining workplaces / Joel
Loow, Bo Johansson, Eira Andersson, and Jan Johansson.
Description: Boca Raton : Taylor & Francis, a CRC title, part of the Taylor &
Francis imprint, a member of the Taylor & Francis Group, the academic
division of T&F Informa, plc, 2018. | Includes bibliographical references.
Identifiers: LCCN 2018030480 | ISBN 9781138092211 (paperback : acid-free
paper)
Subjects: LCSH: Mineral industries—Technological innovations. | Work
environment. | Miners—Employment.
Classification: LCC TN275 .L66 2018 | DDC 622.068/3—dc23
LC record available at https://lccn.loc.gov/2018030480

Visit the Taylor & Francis Web site at
http://www.taylorandfrancis.com

and the CRC Press Web site at
http://www.crcpress.com

Contents

Preface

While the mining industry is often associated with issues relating to safety and health, there have been important improvements. In some countries, the mining industry now has a safety record similar to that of the manufacturing or construction industry. Even so, these topics remain, and will remain, crucial. And parallel to this the mining industry faces the challenge of an ageing workforce, and work that requires a workforce that is more qualified. It is a challenge that is best overcome by attracting young people – and others who normally are not interested in working for the mining industry – that are skilled and competent. Accomplishing this requires a change in how mining workplaces are designed.

This book represents an attempt to contribute to the knowledge and discussion of how to create mining workplaces that are not only ergonomically sound and safe but also attractive. This challenge requires widening of the design. It is not enough that the workplaces that are developed are good in the eyes of the current employees of the mining industry. The workplaces must also meet the needs and wishes of the future workforce, that is, those who currently are not interested in employment in the mining industry. Such workplaces must be designed through their participation. But in addition to this, designers, planners, and other professionals who are responsible for developing and designing the workplaces of the mining industry must have an understanding of the nature of the problems that relate to creating workplaces that are attractive.

To this end, this book reviews theories on workplace attractiveness, research on health and safety in mining, the contribution of technology and work organization to the situation in mining, and so on. It also proposes structured approaches to designing and planning an ergonomic, safe, and attractive mining workplace. It does not, however, offer unequivocal advice. Instead, it aims to provide basic knowledge that we hope will assist in making the right decisions with regard to these topics.

The book builds on experiences that we have gathered through our participation in different mining-related projects. Some of this experience stems from the 1980s when Professor Jan Johansson and colleagues at Luleå University of Technology (LTU) began investigating technology and work organization in LKAB. More recently, our experience builds on our participation in future-oriented – and sometimes visionary – mining and mining-technology development projects. These projects include The Smart Mine of the Future and EU-projects *I²Mine* and *SIMS* (some of these have had specific focus on attractive workplaces). With this book, we hope to share some of the lessons learned so that mining workplaces can be designed that are attractive both to the mining industry's current workforce as well as its future, more diverse and qualified workforce.

Authors

Joel Lööw is a Ph.D. student in human work science at Luleå University of Technology, Sweden. He has an M.Sc. in industrial design engineering with a specialization in production design. His research interests center on issues of the work environment and technology in the mining industry, with a special focus on how these issues can be managed. He has experience working in the EU project, Innovative Technologies and Concepts for the Intelligent Deep Mine of the Future (I^2Mine), and is currently involved in a mine-accident prevention research project and in the Attractive Workplaces work packages of the EU-project, Sustainable Intelligent Mining Systems (SIMS). Lööw is involved in courses on workplace analysis, industrial production environment, production development, and organizational change management for M.Sc. engineering students. He teaches the subject of health and safety to mining engineering students. Lööw has published on the subjects of mining, work environment, technology, and work organization.

Bo Johansson is a retired assistant professor in human work science. He was active at the Department of Business Administration, Technology and Social Sciences, at Luleå University of Technology, Sweden. His research is mainly in the field of work environment management, work environment, and industrial production development in mining and manufacturing business. He has an M.Sc. in mining engineering and a Ph.D. in human work sciences. He has previously worked as a mining engineer and head of a mine planning division at Boliden Minerals. At Luleå University of Technology he has been director of studies and programme coordinator. In 2007, he was awarded the Levi Prize (a national work environment prize) by the Swedish Association of Graduate Engineers.

Eira Andersson is currently working as a quality strategist for equality in Luleå municipality. She was previously an associate senior lecturer in the subject of industrial work environment at the Department of Business Administration, Technology and Social Sciences, at Luleå University of Technology, Sweden. She has an M.Sc. in ergonomic design and production engineering and a Ph.D. in human work science. Her research focuses on industrial relations, occupational health and safety, technology and gender issues in mining production. She has been engaged in several interactive R&D projects together with the Swedish mining industry as well as transnational collaborations, all aiming for attractive, safe, and gender-equal workplaces in mining. In 2013, the Geological Survey of Sweden (SGU) awarded her for significant research contribution to the mining industry, characterized by innovation, sustainability, and societal benefits.

Jan Johansson has been a professor of industrial work environment since 1994. He earned an M.Sc. in industrial management and engineering from Linköping University of Technology in 1975 and a Ph.D. in human work sciences from Luleå University of Technology in 1986. In 1999, Johansson was appointed as Honorary

Visiting Professor at the School of Industrial Relations and Organisational Behaviour, University of New South Wales, Sydney, Australia. Johansson has been a member or the Swedish Research Council (2001–2003) and the Swedish Research Councils Ethical Committee. He has been Deputy Dean at Faculty of Technology and, until recently, he was Head of the Department for Business Administration, Technology and Social Sciences. Johansson's main research interest is work organization and attractive workplaces and is conducted at the Centre of Advanced Mining and Metallurgy at Luleå University of Technology. Johansson has been active in both the Swedish-Polish project, The Smart Mine of the Future, and in the European project, Innovative Technologies and Concepts for the Intelligent Deep Mine of the Future. His current activities include being a work package leader for the Attractive Workplaces work package in the EU-project, Sustainable Intelligent Mining Systems (SIMS). Johansson has more than 200 publications.

1 Modern Mining and Its Challenges

1.1 INTRODUCTION

During the last decade, the mining industry has undergone major developments, which are most evident in improvements in safety. For example, the current safety levels of the Swedish mining industry are almost the same as that of manufacturing or construction industry in Sweden (Swedish Work Environment Authority 2016). At the same time, Elgstrand and Vingård (eds. 2013, p. 6) reported that 'Where reliable national statistics exist, mining is generally the sector having the highest, or among 2–3 highest, rates of occupational fatal accidents and notified occupational diseases'. In other words, notwithstanding significant improvements, work still remains to be done. In addition to this, in many countries, the mining industry faces a different type of challenge: that of ensuring the supply of a qualified workforce in the future. The mining industry is not attractive enough to entice skilled and young people. The problem is two-pronged: (1) the current workforce is ageing and is not being replenished (Hebblewhite 2008; Oldroy 2015) and (2) the changing nature of mining work will require a new set of skills and competences, such as abstract knowledge and symbol interpretation (Abrahamsson and Johansson 2006). Those who possess this knowledge are likely to be found in a population that may be even less interested in employment in mining. Lee (2011, p. 323) framed the problem as:

> A shortage of qualified miners in all types of positions is a critical issue in many countries and regions of the world. During the last decades, as mining declined, the work force was not replaced. … many companies are unable to meet demands because of the severe labor shortage. People currently employed in mining are retiring, and there is a lack of younger people to fill the vacancies.

Hutchings et al. (2011) also indicated the difficulties the mining industry has in attracting employees with appropriate skills. They argued that organizations should give greater attention to strategies that attract and retain skilled employees. To solve the problem, the mining industry employs two primary strategies. On the one hand, the mining industry tends to deal with this issue at a strategic level. Aiming to attract skilled workers, the industry promotes the advantages of working in mining, such as career development, high salaries, and travel opportunities (cf. Randolph 2011). On the other hand, the mining industry is techno-centric and seeks technological solutions to problems (cf. Albanese and McGagh 2011; Hartman and Mutmansky 2002; Lever 2011). Recently, however, the workplace has been recognized as important in facilitating attractive jobs. Lee (2011) identified work organizational strategies for increasing the attractiveness of the industry: avoiding long hours and shifts that do not match modern lifestyles, for example, and upskilling and multi-skilling personnel.

Similarly, Johansson et al. (2010) identified safety, the physical and psychosocial work environment, and social responsibility as important areas for increasing the attractiveness of jobs in the mining industry. Workplace-level interventions can, according to Hedlund (2006, 2007) and Hedlund and Pontén (2006), increase the attractiveness of industrial jobs. Yet, workplace-related changes to increase work attractiveness remain a rare strategy within mining companies and is in general under-investigated. It is here that we see the purpose of this book: it offers guidance for designing more ergonomic, safe, and attractive mining *workplaces.*

The concept of 'attractiveness' (although explored extensively in Chapter 2) needs additional clarification here. In one sense, 'attractiveness' widens the concept of health and safety. In the other, it is a specification of the ergonomics concept. In the first sense, modern methods of accident prevention are holistic and take into consideration the full span of human, organizational, and technological aspects (cf. Harms-Ringdahl 2013). The safety aspect of mining is very important and should always come first. However, a completely safe mine is not necessarily attractive. But we argue that an attractive mining workplace is by necessity safe and that attractiveness can contribute positively to safety. We will try to illustrate why this is, and by doing so, we aim to show how modern safety and accident prevention approaches do not necessarily consider these topics.

Regarding ergonomics, the topics we cover are in some sense all a part of this field. In part, this book exemplifies the application of theories of ergonomics to mining workplaces, but primarily it illustrates how these theories and human-centric design facilitate the creation of attractive mining workplaces. By presenting the concept of 'attractiveness' as a specification of ergonomics, we mean that it gives guidance on which theories and their applications contribute to attractive mining workplaces; as theories of workplace attractiveness are based on theories of ergonomics (cf. Åteg et al. 2004; Hedlund 2007), the factors that contribute to workplace attractiveness represent a specific subset of theories of ergonomics.

The topic of this book is not solely about securing a future workforce, though; like the discipline of ergonomics, it is also very much related to productivity and efficiency (cf. IEA 2017). Attaining ergonomic, safe, and attractive workplaces should be an objective in itself, but there are several other advantages to it as well. These include lower costs, higher productivity, improved quality, and so on. Research has shown that there is a connection between safe and healthy workplaces and those that are productive and efficient. For example, Neumann and Dul (2010) compared 45 scientific studies and found that 95 per cent of them showed a connection between human and system effect: if system performance was poor, then employee well-being was also poor, and if system performance was good, then employee well-being was also good. Another study (ILO 2006) found that there is a correlation between competitiveness and accidents at work. That is, countries with lower levels of occupational accidents are generally more competitive.

Of course, these topics should not be reduced to solely economics. And nor does working with them not incur costs. Health and safety, for example, cannot be attained free of cost: investment is needed to create and maintain healthy and safe workplaces. Our argument (and we are far from the only ones to argue this) is that spending money on health, safety, and related topics now saves money in the future.

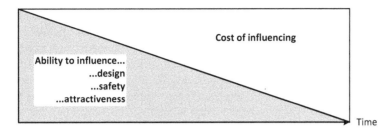

FIGURE 1.1   The ability to influence and cost of influencing design, safety, attractiveness, and so on over time.

According to the American Society of Safety Engineers (cited in Blumenstein et al. 2011), for example, every dollar spent on prevention can save three to six dollars in loss avoidance.

There are other aspects to this as well. It is during planning that the most important decisions regarding work environment and safety are made, that is, when deciding on mining methods, technology, work organization, etc. To be able to influence the aspects relating to ergonomics, safety, and the attractiveness of the workplace, they have to be considered at the early stages of a project. Figure 1.1 illustrates this relationship as it is commonly expressed. One aim of this book is to give an understanding of how to work with these issues early.

To summarize, McPhee (2004) put it clearly. He argued that mining is similar to other heavy industries, and the same principles of design apply in mining as in other industries. While top priority is still assigned to mining disasters and fatalities, the emphasis is changing to include a broader health and safety focus. But, he argued, 'there still appears to be a poor understanding about the contribution of ergonomics to mining, the range of factors that it includes and how these might be addressed', and that the different position of the mining industry is due to 'the different systems that have arisen specifically for mining and that may need to be accommodated through planning and design' (McPhee 2004, p. 297). He saw several emerging issues in ergonomics in mining. For example, there is a change in work practices and a drive for increased efficiency. And some of the practices adopted by the mining industry are at odds with the need to improve the health and safety in general (McPhee 2004). This book takes a special interest in these emerging, as well as traditional, issues and discusses how they can be accommodated through planning and design.

1.1.1   The Position, Purpose, Structure, and Context of the Book

Currently, there are several books on health, safety, ergonomics, and related issues in the mining industry. Some seminal works include Horberry et al. (2011, 2018), Simpson et al. (2009), and Laurence (2011). This book, however, takes a wider perspective. On the one hand, it focuses on the wider issues of attractiveness. As noted, this is not necessarily the same as safety or ergonomics. On the other hand, this book focuses on workplaces and places special attention on workplace-level strategies and practices. In doing so, we try to accomplish two things. First, we give an account of issues – particularly of a social nature – that modern mining is faced with and their nature. Second, we propose two approaches to address these issues operationally and more strategically, respectively. This is not to say that the aspects are unrelated or separate. On the contrary, we hope to show that it is because the issues are what they are that they require the proposed approaches. We do not, however, offer unequivocal advice. The situation in mining, we argue, is context-dependent. And the best alternative in one context may be the worst in another situation. Thus, our strategy is to provide basic knowledge that we hope will help in making the right decisions based on facts regarding the conditions in each unique situation.

The book is structured as follows. In this chapter, we focus on the characteristics of modern mining together with some of the challenges it faces and the present trends. In the Chapter 2, we focus on the concept of attractiveness, and how it relates to other fields of study. Chapter 3 focuses on injuries and ill health in mining, while Chapter 4 focuses on mechanization, automation, and new technology. In these two chapters, we try to show how these issues go beyond traditional 'engineering' problems: that they are problems of a social nature as well. Chapter 5 covers work organizational issues in mining. Special attention is given to how some modern organizational ideas can be adapted to mining, and their role in addressing work organizational issues and related challenges. This constitutes the first part of the book. The second part focuses on how to generally approach and address the issues brought up in the first part; in Chapter 6, we cover some of the principles behind work environment management, while we in Chapter 7 focus on how to work with iterative, user-centric design and planning methods.

This book also represents an attempt to introduce to mining ways of thinking and methods that are common in other industries. In this, the mining industry poses an interesting challenge because it is not solely a process industry, construction industry, and so on. It encompasses all these things. As an example, commentators in the process industry have argued for 'operations centre' concepts. This is not fully possible in the mining industry because it includes aspects from several other industries. The design of mining workplaces must be carried out in way that not only considers all these factors, but that also makes them work together. Additionally, the perspective introduced in this book includes the fact that mines, new mine levels, and so on represent significant investment. They are not realized overnight and can require several years of planning. Therefore, making mining more attractive, safe, and ergonomic in existing workplaces requires as much work as workplaces of the future do. At the same time, it is not enough (and sometimes not even possible) to work with these topics in the operating stages of a mine or development project. These topics require attention from the very first project stage to the very last, which is why these stages are given specific attention in this book.

Finally, in this book we use many examples and experiences from Sweden. On the one hand, we believe that the Swedish and Nordic perspective stands to offer a lot in this context. On the other, there is plenty of Swedish language research that does not reach an international audience. As a result, several successful examples stemming from the Swedish context have not been showcased outside of Sweden. Sweden also has a long tradition of mining with plenty of research looking at its organization, labour, technological development, and so on. Thus, a secondary purpose of this book is to disseminate this knowledge to a wider audience. To this end, we next look at the development of Swedish mining company and how these experiences should be applicable to other mining companies as well.

1.2 HISTORICAL AND MODERN MINING

Mining is a diverse activity. Elgstrand and Vingård (eds. 2013) described the situation in 16 mining countries. At one end of the spectrum, there is the artisanal, small-scale, and sometimes illicit mining in, for example, Ecuador and the Democratic Republic of Congo that persists under poor and dangerous working conditions. At the other end of the spectrum is the highly mechanized and automated mining of, for example, Australia and Sweden. Of course, before becoming modern, high-tech operations, these mines were also characterized by frequent accidents and poor work environments. In this section, we review this type of development in the mining industry. The reason for doing so is that this development has included technical, organizational, administrative, and other changes; some of these are very closely connected to the topics addressed in this book. Through this review, we hope to showcase the breath of possible developments in mining.

1.2.1 THE EVOLUTION OF A SWEDISH MINING COMPANY

We use a Swedish mining company to represent one of the evolutions of Swedish mining. Though the implication of the Swedish context should not be overlooked (for example, in terms of legislation, worker rights, culture, and so on), we hold that the developments themselves are neither necessarily unique to the Swedish context nor to a particular company. That is, we believe that the developments presented here have also occurred in several other companies. A lot can be learned, then, by studying these developments.

Johansson (1986) described work and technology in a Swedish underground iron ore mine between 1957 and 1984 in detail. He noted that in 1957, work in the mine utilized several machines, but the need for heavy manual labour was still significant. (Figure 1.2 depicts drilling around this time.) The degree of mechanization and technological sophistication was asymmetric throughout the mine. For example, the production activities used modern and sometimes semi-automated machines, while development work still utilized older types of machinery involving manual operations. (Figure 1.3 depicts a loader that was used in development work during this period.) The semi-automation consisted in, for example, partial automation of drill rigs where the operator had to feed drill steels to the drill. Work was organized into teams that were responsible for entire production cycles and planning. The company considered itself to have made use of technological development to improve work conditions and to increase productivity. Additionally, due to the demands for labour during this period, they implemented progressive staffing policies. At that time, the director of the company was illustrated as saying:

> In question of salaries, pensions and working hours, our company's miners should be better off than any other comparable group of industrial workers in Sweden ... The development that has led to the rapid improvement of working conditions with regards

FIGURE 1.2 Drilling in development work in the 1950s. (Photo: Börje Rönnberg, courtesy of LKAB.)

FIGURE 1.3 An overhead loader used in the 1950s. It functioned by throwing rocks over the machine into a carriage. It was nicknamed 'the pig' due to the noise it emitted during operation. (Photo: Börje Rönnberg, courtesy of LKAB.)

to spaces, ventilation, illumination, and so on, and to a lighter and less dangerous work, is important. And in the long run, those factors that relate to ... that which can be summarized within the terms of satisfaction and well-being at work, may be even more important.

(Johansson 1986, p. 125, our translation.)

In 1962, salaries in the production activities were high, but those of the developmental work were lower. The technological asymmetries were still present but less pronounced; for example, the developmental work and production activities now used the same kind of loading and hauling machinery. This reduced the requirement of manual labour. Charging had been mechanized in the production cycle. Previously, electrically driven trucks had been used but were now replaced by more efficient diesel-driven trucks. The introduction of diesel brought with it several work environmental issues related to increased noise levels, vibration, and exhaust fumes. (Figure 1.4 depicts loaders and trucks, fitted with cabins, used around this period.) Production groups were broad and included chargers, loaders, drivers, and repairmen. None of the group members were tied to a certain task but rotated between them. A group consisted of 17–28 workers, four of whom were trained to do repairs.

By 1969, the introduction of remote control had begun (although not from control rooms). Mechanization had continuously increased, which entailed investment in expensive machines. To ensure return on the investment, this led to a requirement of full utilization at certain places in the mine. Operators experienced increased psychological pressure as a result. In other areas of the mine, the level of automation and mechanization

FIGURE 1.4   A truck and loader used in the late 1960s. Note the introduction of isolated cabins. (Photo: Börje Rönnberg, courtesy of LKAB.)

remained the same, though with more modern equipment, which above all else meant increased capacities (greater loads for trucks and loaders, for example). Isolated work was more common; in some cases, workers would only meet at shift changes. Planning had moved from the employees to a specialized planning department.

In 1974, many operations were unchanged, but machines had continued to increase in size, capacity, and power (the workers had input in requisitioning this equipment). The new machines represented an improvement of the work environment (though some of these improvements came from enhanced ventilation). (Figure 1.5 depicts remote-controlled tunnel boring machine from around this period.) By that time, however, all machines were diesel-powered. While the production groups still existed at this time, they had been altered. For example, due to changes in the wage system that involved the removal of piece-rates (following a strike), the requirement on co-operation was less pronounced. The foremen experienced a change in responsibility as well as the loss of both freedom and resources. The chief executive officer (CEO) at that time claimed that the self-controlling production groups now required direct input from the foreman. The succeeding CEO held that the non-piece-rate wages were problematic in terms of productivity because it made performance increases more difficult. The responsibility of planning was moved entirely to the production department. This aimed to maximize machine utilization and to ensure certain characteristics of the ore.

By 1978, the company was in deep crisis (see Section 1.2.2), which meant that personnel-related issues came to focus on keeping the company alive, and the company had stunted development. However, the CEO at that time spoke positively of autonomous groups and decentralized decision-making. Work tasks remained mostly unchanged.

FIGURE 1.5   A remote-controlled full-face boring machine used in the 1970s. (Photo: Börje Rönnberg, courtesy of LKAB.)

FIGURE 1.6 Work in a control room in the 2000s. (Photo: Lena Abrahamsson.)

In 1985, the tide had turned once again and the company produced good results. The work system remained much the same, but there had been several work environmental improvements in addition to capacity advances. Electrically driven machines were reintroduced and all employees were trained regarding quality.

Abrahamsson and Johansson (2006) provides an account of the situation of the company in 2005. The principal change that occurred during this time was the increased automation and remote control. This meant moving the control of certain operations to control rooms. Teams of operators oversaw the processes, but each operator had a specific task (i.e. an operator was a driller, a loader, and so on). There was also spatial divisions between the operators as certain operations were performed from certain control rooms. (Figure 1.6 depicts control room work around this period.) Table 1.1 summarizes the entire technological development.

From this we conclude that developments in mining, be it to increase productivity or due to workforce requirements, has included both organizational and technological developments. This review should also show that that these developments depend on each other and that one development hinders or facilitates another.

1.2.2 THE STEEL CRISIS AND PRODUCTIVITY GAINS

The steel crisis persisted from around 1974 to 1983. It drastically decreased iron ore prices, leaving most iron ore mines struggling for survival. Few iron-producing countries were unaffected. It is interesting to us because it shows how some organizational practices that are usually associated with attractive and ergonomic workplaces can also increase productivity – in this case, to the extent of being the difference between continued survival and bankruptcy.

Galdón-Sánchez and Schmitz Jr. (2002) showed that the iron-ore mines that survived the crisis did so mainly by increasing their productivity, which had remained at the same level for long. After the crisis, few, if any, mines managed to reach the pre-crisis production volumes but did manage a *productivity* increase of around 100 per cent. Galdón-Sánchez and Schmitz Jr. (2002, p. 1224) argued that this increase was not achieved through the closing of unproductive mines, shifts in type of iron ore produced, or the introduction of new technology, but was 'primarily driven

TABLE 1.1

The Technological Development of a Swedish Mining Company

Process Type	Work Task	Technology Level			
		1957	1969	1985	2005
Material extraction	Drilling	2/3	3/3	3/3	3/6
	Charging	1/1	2[a]/2[a]	2[a]/2[a]	2/2
	Blasting	1/5	5/5	5/5	5/5
Transport	Loading	3/3	3/3	3/3	3/3
	Hauling/dumping	3[a]/3	3/3	3/3	–/5
	Chute loading	–/4	–/4[b]	–/5	–/5
	Dumping	–/3	–/4[b]	–/6	–/6
Processing	Crushing	–/4	–/6	–/6	–/6
Hoisting	Skip loading	–/4	–/6	–/6	–/6
	Skip hoisting	–/6	–/6	–/6	–/6

Source: Based on Abrahamsson and Johansson (2006).

Notes: The figures denote technological level.

1, manual work; 2, motor manual work; 3, machine work; 4, operating work; 5, remote control; 6, automated work.

In the 'x/y' notation, 'x' refers to the development unit operations, and 'y' refers to production unit operations.

[a] Some activities remained that are classified as 1.

[b] Some activities remained that are classified as 3.

by continuing mines, producing the same product, and using existing technology'. Tilton (2014) argued, on the other hand, that low prices motivate mining companies to adapt new technology and work practices, and that this was the case for US mines during the crisis. Still, while noting that incremental advances in the form of bigger and more powerful machines as well as the increased application of mathematical and computerized planning tools in iron ore mining increase productivity, Bartos (2007) identified no revolutionary technology that could explain the gains in iron ore mining productivity. Instead, Schmitz Jr. (2005) argued that the main reason why Canadian and US iron-ore mines were able to survive the crisis was due to changes in the work organization. This included widening operator roles to include the operation of several different machines and repairs (previously operators were more limited in their roles). As will be seen, this type of change can be in line with attractive work practices; the consideration of these issues are, or should be, close to the operation of mines in general.

1.2.3 SOME CHARACTERISTICS OF MODERN MINING

The last part of our review of modern and historic mining focuses on its characteristics with regards to the development in employee education, professional roles and occupations, and some other aspects of its labour. Again, we look at Sweden.

Figures compiled by Johansson (1986) show that in the early 1950s, around 10 per cent of all employees of a Swedish mining company were white-collar workers. Around 10 years later, this figure was around 15 per cent, and in the early 1960s, it increased to around 20 per cent. In the late 1970s and into the 1980s, around 22 per cent of the workforce consisted of white-collar workers (figures exist up until 1984; after that it is difficult to find this type of statistics). Beyond this period, we use educational levels in the Swedish mining industry (see Table 1.2). While these numbers capture an increase in national educational levels, it is also important to note that the lowest levels of education have decreased considerably: an increasing number of jobs now require at least some upper secondary education and the majority of jobs requires a full three-year education. On the tertiary level, education to bachelor level or higher has increased almost four-fold. All in all, the educational level is increasing.

Another important point of consideration is the structure of the workforce. Table 1.3 presents figures from the Swedish mining industry in 2014. Note that the table is not intended as a detailed description of the structure of the workforce. For example, the table presents the ten most common jobs within the industry, instead of the ten most common jobs within each sector. Rather, its purpose is to illustrate a few notable points. Although the second most common job in the industry is operators driving machinery, it appears as a rather uncommon profession in the actual mines. In fact, in the mines, technology and natural science specialists make up a considerable part of the workforce. In general, low-skilled labour is considerably rare. However, these figures do not contain an important factor: contractors. Contractors have come to be responsible for a considerable amount of the hours worked within the industry, but as a rule, are not included in statistics relating to the mining industry. Recently however, they have received more attention. For example, the Geological Survey of Sweden (2016) included contractors in their latest publication. These estimations show that around 12 per cent of all employees in the mining industry are contractors.

TABLE 1.2

Educational Levels in the Swedish Mining Industry, 1995 and 2014

Education Level	1995 (%)	2014 (%)	Change[a] (%)
Primary school	34.7	11.5	−23.2
Upper secondary school, up to 2 years	42.6	31.7	−10.9
Upper secondary school, 3 years	13.2	35.0	+21.8
Tertiary education, less than 3 years	5.8	8.9	+3.0
Tertiary education, 3 years or more including doctoral studies	3.6	12.7	+9.1
No data	0.2	0.3	−

Source: Data from Statistics Sweden (1996, 2016a).

Note: The classification system of both the education level and industry changed between 1995 and 2014, but in general give a clear indication of the changes that occurred.

[a] The change in percentage points; a minus signifies a decrease and a plus an increase.

TABLE 1.3

The Ten Most Common Types of Jobs in the Swedish Mining Industry 2014

	Mining of non-Ferrous Metal Ores	Mining of Iron Ores	Quarrying of Limestone, Chalk, etc.	Operation of Gravel and Sand Pits	Total
Stationary plant and machine operators	673 (34.5)	1211 (33.2)	183 (67.0)	232 (17.2)	2317 (31.2)
Driving and mobile plant operator trades	120 (6.2)	162 (4.4)	25 (9.2)	593 (44.1)	926 (12.5)
Metal, machinery, and related trades	78 (4.0)	613 (16.8)	14 (5.1)	84 (6.2)	792 (10.7)
Technology professionals	262 (13.4)	429 (11.7)	14 (5.1)	60 (4.5)	774 (10.4)
Natural sciences and technology specialists	165 (8.5)	257 (7.0)	–	6 (0.4)	428 (5.8)
Other professions	84 (4.3)	69 (1.9)	32 (11.7)	72 (5.4)	351 (4.7)
Electrical and electronic trades	77 (4.0)	249 (6.8)	–	6 (0.4)	332 (4.5)
Managers of IT, logistics, R&D, engineering, construction, etc.	85 (4.4)	108 (3.0)	5 (1.8)	72 (5.4)	273 (3.7)
Recycling collectors, newspaper distributors, and other service workers	–	195 (5.3)	–	–	195 (2.6)
Office assistants and secretaries	41 (2.1)	62 (1.7)	–	69 (5.1)	172 (2.3)
Total	1585 (21.3)	3355 (45.1)	273 (3.7)	1194 (16.1)	6560 (88.2)

Source: Data from Statistics Sweden (2016b).

Notes: Only the professions with the ten most employees, as calculated by industry totals, are included. Only sectors with at least 100 employees are presented, but all employees of the industry have factored in the calculations.

The number in parenthesis is the percentage of all employees in that sector, but for totals, this is the percentage of all industry employees.

Other sources put the hours worked by contractors at around 40 per cent in some cases (Nygren 2016).

Another development is the contraction of the industry in terms of the number of employees and places of work. In the 1950s, there were more than 100 places of work and between 12,000 and 15,000 employees. In the 2010s, there were fewer than 20 places of work and below 8,000 employees (Geological Survey of Sweden 2016). This means that while the number of mines is lower, those that remain have grown in size. At the same time, mines are more productive than ever (with regards to tons produced per man-hour), even if production volumes are lower than before the steel crisis.

With this brief statistical portrait we want to illustrate the transformation of the mining industry towards more advanced operations that require high-skilled labour and the increasing size of individual mining companies. Altogether, including the utilization of contractors, this creates more complex organizations. The change in structure and increasing complexity require different approaches to questions of labour, ergonomics, and safety than have traditionally been applied.

1.3   FUTURE TRENDS AND CHALLENGES FOR THE MINING INDUSTRY

The mining industry is subject to change. While managing changes, the world's metal mining industry also faces a number of external challenges that must be managed wisely. Some of the challenges include international competition that forces the industry to further rationalize, which requires both new technologies and organizational forms; the impact of the Paris agreement that will increase costs for treating greenhouse gases and acquisition of emission allowances; ore that will be mined from gradually deeper deposits with increased rock stability problems; and that it will become increasingly more difficult to recruit skilled workers to mines that will more often be located far from larger communities (Abrahamsson et al. 2009). At the core of this book is the conviction that these problems must be addressed with a so-called sociotechnical perspective that covers the entire mining and minerals value-adding chain. The mining industry needs a revised mental image of itself based on new technology. It also needs a modern work organization that supports both high productivity and good working conditions. This is necessitated, we argue, by the nature of the challenges that the mining industry faces. To this end, in this section, we attempt to capture and discuss future developments in 14 trends that will affect the future mining industry (although some of these trends are already affecting the industry). This overview takes its starting point in Abrahamsson et al. (2009) but has been condensed and refined in light of recent developments in the industry and current mining research projects (for example, *I²Mine*, *SIMS*, and *Smart Mine of the Future*).

1. Future mining will be shaped in a context where it is necessary to produce at costs that are determined in international competition. The prices of metals and minerals are set by the market, but in the long term, there is little doubt that the demand is increasing. Large nations like China, India, Indonesia, Brazil, and the whole African continent will require a larger amount, and this will lead to the opening of new mines. The difference between the annual 'per capita' consumption of these countries and that of Western Europe can be more than ten-fold.

2. Production conditions will be characterized by the fact that nearby and easily accessible ore will be mined first. New ore will become more distant and found at greater depth. Large ore reserves are located under the sea, and it is likely that mining and offshore companies will develop new technology to extract those. In both cases, production costs will increase. In regard to these first two points, Tilton (2010) argued that mineral depletion is not a

question of physical depletion but rather one of cost becoming too high to justify exploitation. That is, the cheaper that each ton of mineral can be produced, then the more minerals there are 'available'. This is achieved through both technological and organizational measures.

3. Mining depths increase and bring new stability problems. The role of rock mechanics in the design of layouts, cutting sequences, strata stabilization, roof bolting, etc., must be a central issue in the future. However, organizational aspects are also important. For example, rescue operations become more complex with increasing depth and will place more requirements on organizational aspects. On the technical side, techniques such as full-face drilling and cutting may prove interesting from a safety perspective. Production drilling and blasting for controlled fragmentation are crucial operations in the ore mining cycle. Improvements in these operations can create significant positive impact on subsequent operations.

4. Mechanical rock fragmentation opens up many possibilities for automation. A controlled fine-grained fragmentation allows the ore to be processed and transported in a rational way, for example, by pumping. Additionally, this type of change of technology has previously facilitated new forms of organization (cf. Trist and Bamforth 1951; see also Chapter 2).

5. The circular economy affects both energy consumption and management of emissions. The discussion of energy consumption is closely linked to global warming and carbon emissions. An improvement of energy efficiency will increase economic profitability as well as reduce environmental impact. Many components affect total energy consumption. One frequently discussed concept is underground pre-concentration (in situ), which directly affects the energy-consuming activities of hoisting and milling. A change from, for example, diesel power to battery power can also be a strong alternative for the mining industry because, in some cases, there are only minor technical problems with such a change while the environmental and economic benefits are many. Full-scale tests have indicated the potential for battery power, but more research and international collaboration are needed. New technology of this kind is, however, not automatically accepted by its intended user. If not carefully managed, there is often resistance. Additionally, technology such as in situ pre-concentration will affect the working environment. In both cases, attention has to be paid to both the socio-organizational and technical aspects.

6. The discussion of waste management is focused around leaving behind as few footprints in nature as possible; for example, toxic substances must not be left that leak out into nature, and the landscape should be restored as far as possible. While this issue does not relate that strongly to traditional views of the working environment, it is important for the image of mining and thus the attractiveness of the industry. Additionally, some of the techniques employed to deal with these issues, such as in situ mining, can introduce new environmental, health, and safety risks.

7. The environmental debate also includes a discussion on the mining industry's social responsibility for the welfare of the local community. In addition

to preserving the environment, they are obligated to build a strong technical and social infrastructure that ensures the survival of society after mining has ceased. The issue of what a mine leaves behind includes the health of the denizens of the local community and has clear connection to the working environment (see Horberry et al. 2013).

8. Health and safety are very high on the agenda (Ventyx 2012 identified ensuring workforce safety as the top challenge for global mining companies) and is a strong driving force behind the ideas of automation. In underground mining, the basic and first solution must be to distance miners from the physical mining front and locate them in a safer environment. Remote control, automation, and new mining techniques are major challenges and possibilities, but it is also a matter of relevant education, rules, and good practices. In the vision of the safe mine, there is production with zero entry for employees, where personnel are located in control rooms above ground and remotely monitor and control the different operations (see Bäckblom et al. 2010).

9. As noted, in many countries there is a lack of skilled personnel, including miners and mining engineers. Young people are currently not particularly interested in working in an industry that is often located far from larger communities. A further challenge is to break the unequal gender balance that exists in most mines. It is not just about recruiting more women, but rather to challenge the prevailing 'macho culture' in order to create a safer and more productive environment. To recruit the right workforce, mining companies need to improve their knowledge of both attractive and repelling work features. Issues such as these will inevitably involve various work environment issues.

10. There are clear signs that digitalization attempts such as the Industry 4.0 will become more apparent and even common in mining workplaces. This development requires reflection and consideration so that these technologies do not create more problems than they solve; it is necessary to analyse and understand the relationship between new technology, working conditions, qualifications, identity, gender, and human–machine interaction with a focus on integrity issues (see Johansson et al. 2017).

11. Extended business and open collaboration are two concepts where virtual reality (VR) technology can be used to link production functions (such as planning, mining, maintenance, logistics, purchasing, and for coordination of external contractors, suppliers, customers, etc.) to a production flow or value-adding chain, where all share the same goal and everyone sees the same 'whole'. Common visualization of problems and opportunities in the system allows everyone to optimize the whole chain rather than sub-optimizing parts. This, like most technology, creates new conditions for work organization. For example, the nature of social contact might be significantly changed.

12. Production centres create new professional roles, and the mining industry will experience an increasing degree of remote control from production centres and collaborative visualization rooms. These may be located in nearby

communities or further away (other continents), where the operators have monitoring and coordinating activities across the value chain (facilitated by some of the technology described above). Their jobs will have changed character towards service work, and the new tasks require different kinds of skills. In addition, to deal with advanced information technology, miners have to interact with different specialist teams located all over the world.

13. Modern technology has created a new type of work in terms of competencies, knowledge, and workload. There is an emerging, and in many aspects already evident, knowledge transformation: from the old, obsolete physical and tacit knowledge, and skills (e.g. the ability to 'read the rock') to something new that can be described as abstract knowledge. The new forms of work in mines have less need for the traditional mining competencies, attitudes, and ideals. The traditional workplace culture and 'macho style' is challenged; workers have to find new ways to learn and develop a workplace culture more attuned to a new type of worker identity and masculinity (Abrahamsson and Johansson 2006).

14. Mines of the future will have smaller workforces and will need a different kind of model for work organization than today. Mining companies will gradually become flatter and leaner organizations with multi-skilled workers who can operate in several areas and functions within the company.

There are, of course, other important areas of development, but those above are among the most important from a long-term strategic and sustainability point of view. To summarize, the mining industry is facing major challenges that require extensive and immediate action. Problems interact with each other in such a way that the solutions require a holistic perspective where both technical and social factors are taken into consideration. There is a need for a new and modern vision for the whole industry, based on a sociotechnical approach, that covers the entire mining and minerals value-adding chain, including environmental issues.

1.3.1 A Vision of Future Mining

A vision for future mining, which especially considers the trends above, was presented by Johansson and Johansson (2014). It is summarized here. The vision describes future deep metal mines as planning and cooperation successes. This type of mine would utilize automated and flexible mining systems based on drill-and-blast technology for ore fragmentation and continuous mechanical fragmentation (e.g. roadheaders) for development work. It would largely satisfy the requirements for zero-entry mining. Such a mine would use automated mining methods that make it possible to continuously produce desired ore qualities and quantities based on customer demand (providing a significant competitive advantage). This mining system would have dramatically reduced the prevailing and traditional use of storing and stacking mined ore and would allow for the practice of 'lean mining'. The vision describes automated mining technology as being able to reduce costs for underground development work by 50 per cent and labour costs by even more. This would make it possible for companies to make large investments in new technology and personnel competence while still being highly profitable. New mines started under this vision would be exclusively underground. A green mining philosophy of in situ mining would be applied, making most mining activities almost invisible. For example, waste material would be directly used for backfill after recovering the metal content. This visionary mine would have been developed using an iterative planning and design methodology that reduces initial design errors and reflects 'Safety First' in every design decision. All activities would have been risk assessed and physical work properly simulated, evaluated, and approved before being conducted. The mine would have been designed from the start with sociotechnical principles (e.g. work organization based on production teams and broad professional skills among management and miners) and automation in mind. It would feature an impressive information and decision system based on sensor technology and production analysis. This would allow for proactive steering and controlling of the production process, increasing quality and production availability and stability, as well as making miners' work interesting and challenging. The new deep metal mine would make use of remote operations centres designed to promote cooperation and creative problem solving in multi-skilled teams (made up of people of different ages, experience, gender, competence, background, and so on). Mining work would have become attractive not only because of high wages, but because of interesting work with good possibilities for personal and profession development in a safe and sound working environment.

While the fulfilment of this vision would require development and innovation on several fronts, not least technologically, we hold that this book provides some of the tools and knowledge required to reach this goal. It shows that some of the change required can also be enacted without technological innovations.

1.4   CONCLUSIONS: THE NEED TO MAKE MINING MORE ERGONOMIC, SAFE, AND ATTRACTIVE

We conclude this chapter by stating that future mine work will change and is to some extent already changing. Subsequently, the future mineworker is different from today. Already, the average mineworker is more educated, partly owing to the changed nature of mining work. Additionally, further changes to the technology of mining will require a different set of skills and competencies. On the one hand, not only are the vast majority of people currently not interested in working in the mining industry, but future potential mineworkers may be even less interested. It is not enough to only rely on technology to solve this issue; the nature of mining work has to change too through organizational interventions as well as through the redesign of its workplaces. On the other hand, the technology of future mining will affect the working environment in a variety of different ways – not all of them positive. Looking strictly to technology to overcome the challenges engendered by technology will not be enough. Again, organizational factors as well as those relating to the design of mining workplaces will have to be considered. Safety is the traditional focus of this consideration, and while this is imperative, it is only one part of the required holistic approach; both safety and technology must be considered in tandem with ergonomics and workplace attractiveness. This is not only a question of securing a future labour supply; it is also a question of effective and efficient production. The social requirements on technology, work organization, and so on also tend to be symbiotic with productivity. However, these requirements and relationships (not only between productivity and social issues but also of attractiveness, ergonomics, and safety) are complex. This complexity is overcome initially through a better insight and understanding of the issues at hand. It is here that this book has its clear contribution; while we do not claim that it represents the definitive answer to these questions, we do hold that it provides the reader with both insight and understanding as well as systematic approaches to work with these issues.

2 Attractive Mining Work

2.1 THE THEORETICAL AND HISTORICAL ROOTS OF ATTRACTIVE WORK

This chapter deals with the nature of attractive mining work and provides the theoretical basis on which our subsequent discussions will be based. Here, we are concerned with mainly two theories of workplace attractiveness. We hold that they are the result of several research contributions within the areas of industrial sociology and psychology, engineering, and similar fields – often referred to collectively as ergonomics (alternatively, human factors) or human work science. A brief review of the three most important schools of thought within these fields is required to facilitate an understanding of the theories of attractiveness. These are the three theoretical bases that most of the work organization theories rely on, including the mining industry: scientific management, human relations, and the sociotechnical school.

2.1.1 SCIENTIFIC MANAGEMENT

Scientific management – also known as Taylorism or the rationalization movement – was developed by the American engineer Frederick Winslow Taylor (e.g. Taylor 1911) at the turn of the twentieth century. He carried out the first time-and-motion studies in a machine shop in the United States, but the mining industry quickly adopted Taylor's ideas. The ideas can be summarized in two theses: (1) there is one best way and (2) the right man in the right place.

It is Taylor's first thesis that has the greatest effect on the design of work organization. Taylor based it on the assumption that professional knowledge is something that craftsmen develop over generations. The difficulty for Taylor lay in accessing this knowledge in order to develop and apply it in a systematic manner. He recommended a way for how this 'best method' should be established and spread throughout the company:

> The managers assume, for instance, the burden of gathering together all of the traditional knowledge which in the past has been possessed by the workmen and then of classifying, tabulating, and reducing this knowledge to rules, laws, and formulæ which are immensely helpful to the workmen in doing their daily work. In addition … the management take on … other … duties …
>
> … They develop a science for each element of a man's work, which replaces the old rule-of-thumb method.
>
> … They scientifically select and then train, teach, and develop the workman, whereas in the past he chose his own work and trained himself as best he could.

... They heartily cooperate with the men so as to insure all of the work being done in accordance with the principles of the science which has been developed.

... The management take over all work for which they are better fitted than the workmen, while in the past almost all of the work and the greater part of the responsibility were thrown upon the men.

It is this combination of the initiative of the workmen, coupled with the new types of work done by the management, that makes scientific management so much more efficient than the old plan.

(Taylor 1913, pp. 36–37)

The real breakthrough for Taylor's ideas came with the methods–time measurement (MTM) method, a system whereby each manual work operation could be broken down into its basic motions. Each basic motion was assigned a set time using a special unit of time. Using the MTM system, it was then possible to design and determine the time and composition of work theoretically.

In his second principle, the right man in the right place, Taylor emphasized the importance of finding rational selection methods. He described the procedure for selecting workers at a steel company to participate in an attempt to load 47 tons of pig iron a day instead of the previous 12.5 tons:

Our first step was to find the proper workman ... We therefore carefully watched and studied these 75 men for three or four days, at the end of which time we had picked out four men who appeared to be physically able to handle pig iron at the rate of 47 tons per day. A careful study was then made of each of these men. We looked up their history as far back as practicable and thorough inquiries were made as to the character, habits, and the ambition of each of them. Finally we selected one from among the four as the most likely man to start with.

(Taylor 1913, pp. 43–44)

Taylor's methods have since developed within the framework of what became industrial psychology. One problem with the Taylorist system was the high costs associated with supervision and checking. The division of labour required in-depth planning, which was an expensive process. Quality control was another costly stage; Taylor recommended a double-control system of supervisors overseeing the workers and senior supervisors overseeing the supervisors. Taylor's system was not founded on placing trust in the workers. He believed that loyalty was the route to success and the guiding industrial principle. At the same time, he was aware that in order for his ideas to be put into practice, he needed the cooperation of his workers. It was therefore necessary to introduce a system that used studies to reward those who completed their tasks within the allotted time. Taylor therefore devised a number of piecework pay systems.

In hindsight, it is easy to criticize Taylorism for its inhuman view of people, and such criticism is justified. However, Taylor's system of rational production also paved the way for significant increases in productivity that benefited the whole of society in terms of greater material prosperity through both higher wages and a wider range of goods at lower prices. Taylorism entered Swedish mining around the 1950s (see Eriksson 1991).

2.1.2 THE HUMAN RELATIONS SCHOOL

The human relations school has its basis in the classic Hawthorne experiment carried out by Professor Elton Mayo at the Western Electrics Hawthorne Works in the outskirts of Chicago from 1924 until 1932 (e.g. Mayo 1945). The background to this study was the growing dissatisfaction with the conditions at the factory, despite employees being treated relatively well; for example, there were progressive pension and health insurance welfare arrangements. The company therefore contacted the American National Academy of Sciences to ask for help.

The outcome of the human relations school studies was that the importance of social relationships was highlighted along with the fact that work is a group activity. Mayo emphasized that the social world of the adult focuses primarily on work. The worker's attitude towards work and their effectiveness at work are determined by social requirements placed on them both within and outside the company. Informal groups in the workplace exercise a significant degree of social control over the habits and attitudes of the individual worker. The need for recognition and security and the sense of belonging to a group have a greater impact on work morale and productivity than the material conditions in the workplace. The human relations school's emphasis on the social dimension has a generality that extends far beyond the industrial sector.

However, it would be some time before the work organization of the human relations school experienced any significant breakthrough; the Taylorist concept, with its strict control and its far-reaching division of labour, remained dominant. Its influence was at a more ideological level, where the human relations school argued for a more humane view of people in stark contrast to Taylorism's more mechanical view. One visible practical consequence was the introduction of personnel and human relation departments at companies. Group piecework arrangements and labour management committees were two other attempts to implant the positive group norms of the human relations school into the prevailing work organization.

2.1.3 THE SOCIOTECHNICAL SCHOOL

The sociotechnical school in a sense can be viewed as a combination of the human relations school and scientific management. The human relations school provides an analysis of the social system, while the scientific management school's rational production technology constitutes the groundwork. Added to this is a system-theoretical approach, that is, the idea that an organization both characterizes and is characterized by its surroundings.

The theory of sociotechnical organization was developed in the 1950s at the Tavistock Institute of Human Relations, and it is based on an analysis of the introduction of new technology in British coalmines (e.g. Trist and Bamforth 1951). Here, mining had been mechanized during the inter-war period, replacing the earlier manual system with the semi-automated longwall method. However, increases in productivity were not as high as had been expected, and an increasing number of workers were leaving the mines despite higher pay and better working conditions.

At this point, researchers from the Tavistock Institute were called in to investigate the problem.

The introduction of new semi-automated technology required a work organization that was more like that of a small factory department. The work was divided into a series of sub-operations that followed each other in a set order over three shifts of 7.5 h each. The shift crew was spread out geographically over an area that was 200 m long and 2 m wide. The different shift crews did not meet each other, and coordinating the new organization required staff management. The researchers concluded that the technical dependence between the different shift crews was not supported by a social system that dealt with the whole process, and that there was a need for integration. Instead, the work organization had the opposite effect: contributing to further fragmentation of the production system. Later on, the Tavistock researchers came into contact with a mine where the longwall method worked very well, where productivity was higher than at other comparable mines. The difference was that the workers themselves had created a work organization based on broader work roles with work rotation both within and between the shift crews. They had succeeded in creating a social system that worked in harmony with the technical system; this new system involved a high degree of autonomy. The researchers had discovered the autonomous group. Organizational choice was possible within the framework of a set technique.

The autonomous group that the Tavistock researchers described did not attract any immediate imitator. Some experimental work was carried out in the Indian textile industry, but it remained relatively unknown in the Nordic nations. However, in Norway, a series of trials with autonomous groups began through collaboration between the Norwegian Trade Union Confederation and the Norwegian Employers' Federation. The researcher Einar Thorsrud was appointed to lead these trials, and cooperation with the Tavistock Institute was established. The aim was to develop a new work organization with more meaningful work content. This content was specified through six psychological requirements of productive activity (Thorsrud and Emery 1969):

1. the need for work to involve more than just endurance: work should involve a certain amount of variation, although this does not mean that new things have to happen all the time;
2. the need to learn something from the work and to continue learning;
3. the need to make decisions, at least within the limited area that the individual calls his or her own;
4. the need for status – at least a certain degree of human understanding and respect in the workplace;
5. the need to see a connection between the work and the surrounding world – at least to see a certain correlation between the work performed and what is considered useful and valuable; and
6. the need to see that work can be combined with a hope for the future, without this necessarily having to include promotion.

The results from the Norwegian trials were promising and showed that a changed work organization with increased worker influence could lead to greater productivity and improved job satisfaction. Inspired by the results from the Norwegian trials, IF Metall, the Swedish Metal Workers' Union (which organizes miners), formulated its concept of 'The Good Work' in form of nine principles for how it can be attained (Metall 1985): (1) job security; (2) a fair share of production earnings; (3) co-determination in the company; (4) a work organization for cooperation; (5) professional know-how in all work; (6) training as part of the job; (7) working hours based on social demands; (8) equality at the work place; and (9) a working environment without risk to health and safety. When comparing Metall's 'The Good Work' to the sociotechnical approach, it becomes evident that they have expanded the perspective from the workplace to conditions on the labour market. And as is shown below, many of the requirements of attractive work take a similar form.

2.2 WHAT IS ATTRACTIVE WORK?

With the theories behind scientific management, human relations, and the sociotechnical school in mind, the connection between productivity, work organization, and workplace attractiveness should be clearer. Expanding on this, in this section we discuss the specifics of attractive work, and the workplaces through which it is facilitated. As will be seen, the connection between these fields is strong though requires some changed perspectives.

The field of attractive work is quite limited and relatively new but has clear connections with older and well-established fields of research on, for example, work organization, work motivation, and work environment (as was reviewed above). While there is no single accepted definition of attractive work, work that is considered attractive is, in essence, work that people want to have and enjoy doing. To this end, Hedlund (2007) proposed the typology presented in Figure 2.1. The model

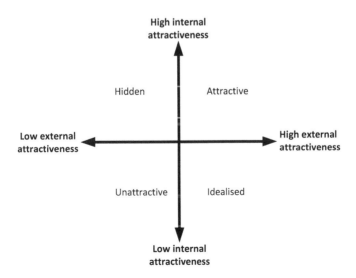

FIGURE 2.1   The two dimensions of the attractiveness of work. (Based on Hedlund 2007.)

has two dimensions that correspond to the two perspectives of work attractiveness: the internal and external perspective. The internal perspective represents a person who is already employed in the job. In this view, the job is attractive if this person wants to keep their job (i.e. the person likes their job) and is unattractive if this person does not want to keep it (i.e. the person dislikes their job). The external perspective represents a person who is not employed in the job in question, usually someone who is looking for a job or is thought of as intended for the job. Here, the job is attractive if the person in question wants the job (i.e. considers it viable) and unattractive if they do not want it (i.e. does not consider it viable). This gives rise to a quadrant that classifies a job into one of four categories: attractive, idealized, hidden, and unattractive.

People take unattractive jobs out of necessity; for example, when there are no other jobs available. In Blauner's (1964) view, people doing unattractive work are most likely alienated (see Chapter 4). Unattractive work should be characterized by high turnover rates, similar to what Blauner found with the workers in the US car industry in the 1940s and 1950s. Occupational diseases caused by organizational and psycho-social factors should also characterize this type of job.

Hidden jobs are essentially 'good jobs' but that only few know about or consider attractive. This may be a problem of communication: people simply do not know of the job. It might be that it is a job in a small sector and, as such, does not get coverage in media. It may also be that the job is located in a remote location, which hinders a sort of 'natural dissemination' of the characteristics of the job. Alternatively, hidden jobs may be more of an 'image problem'. For example, a hidden job may be a job that is in a sector that is normally characterized as containing unattractive jobs, or it may struggle with its historical roots (e.g. the job used to be unattractive but is not so anymore). Hidden jobs can be more difficult to change because it could also entail changing norms. For example, mining may be viewed as unsuitable for women or certain lifestyles. Thus, changing norms is as important as changing the image of the job. For hidden jobs, then, it is seldom the work itself that needs changing but rather the image of it or of how it is communicated.

Idealized jobs are those that are *considered* good but in reality are not. Such jobs can only be improved by changing the work itself. Idealized jobs are not necessarily characterized by high turnover rates like unattractive jobs are, but they could have quite high numbers of occupational diseases due to organizational and psycho-social issues. In a sense, this may be because norms coerce people into the job, and the loss of status associated with the job might be too great and discourage employees from quitting. There is also the case where the communicated image of the job is different from the actual job. In this case, the job should be characterized by high turnover rates. It is also unlikely it will remain idealized for long until turning into an unattractive job.

Attractive jobs are ideal. They are jobs that people want to have and like to be in. In terms of attractiveness, they do not require much changing but need to be maintained. This can include keeping equipment updated and relevant and offering benefits that suit the current and local context. In other words, whatever the situation, knowledge regarding attractive work is important.

2.2.1 CHARACTERISTICS OF ATTRACTIVE WORK

Beyond the classification of jobs into categories of attractiveness, some research has gone into defining the actual characteristics of attractive work and the workplaces in which such work is performed. The results are available in several publications (e.g. Åteg 2006; Åteg et al. 2004; Hedlund 2006, 2007; Hedlund et al. 2010a, b). Åteg et al. (2004) contains the most extensive summary and is the version we have adapted here. The purpose of the section is to use the model to discuss and analyse what contributes to attractive workplaces. Later, we use the model to discuss the potential of mining work to be attractive.

Åteg et al. (2004) suggested that work attractiveness consists of three categories: work content, work satisfaction, and work conditions. These categories are made up of 22 dimensions and each dimension is made up of several qualities that specifies the dimensions (see Table 2.1). The model is general and variable depending on different individuals. For example, one individual may not consider certain dimensions as contributing to attractiveness whereas another would. The model is also inclusive: we believe that there should not be many dimensions that contribute to work attractiveness left out of the model. However, aspects such as the relative importance of each factor depend on the individual.

2.2.1.1 Work Content

Åteg et al. (2004) argued that within attractive work content, *working speed* refers to the tempo at which the work is carried out. The work pace has to encompass *intensity* as well as *calm periods* and *pauses* to contribute to attractiveness, that is, periods of high tempo as well as those of lower tempo that give opportunity for reflection and recovery. Pauses include both longer and shorter breaks, such as lunch and bathroom breaks.

Physical activity refers to the physical activities of the work task. These activities have to represent a healthy strain so that they do not lead to ill health or injuries. Work tasks should also include *spatial movement* to be attractive, for example, moving from one machine to another.

Familiarity means ensuring that the employee *knows what to do* and *what to expect*. For example, it would entail making sure that the employee knows what tasks shall be performed, how to perform a certain task, and what each day will include.

The work content should give *freedom of action*. This entails the ability for the employee to *organize and control their own* as well as *others' work*. This can be influencing the planning of one's own work or that of others.

Attractive work should include *practical work*. This includes the qualities of *working with one's hands* and *creativity*. Working with one's hands entails, for example, breaking from working only with computers, control systems, and computer-controlled machines to work instead with handheld machines or tools. Creativity refers to the act of creating (though not manufacturing or producing). Åteg et al. (2004) took as an example a chair: manufacturing 100 identical chairs is not creative work but constructing that first chair is.

TABLE 2.1

A Model of the Dimensions and Categories that Influence Work Attractiveness

Work Conditions	Work Content	Work Satisfaction
Adequate equipment	*Working speed*	*Sought after*
Modern	Intensity	Coveted skills
Good quality	Calm periods	Important work
High productivity	Pauses	Feeling needed
Healthy workload	*Familiarity*	*Acknowledgement*
Working hours	Knows what to do	Internal
Predetermined work hours	Know what to expect	External
Ability to influence	*Physical activity*	*Status*
Physical work environment	Healthy load	Pride
Locales	Spatial movement	Success
Furnishing	*Freedom of action*	Professional identity
Air quality	Organize own and others' work	*Stimulation*
Sound levels	Control others and others' work	Challenge
Cleanliness	*Practical work*	Development
Leadership	Working with one's hand	Interest
Trust	Creativity	*Results*
Communication	*Mental work*	Direct and visible results
Information	In the work task	Feeling of connection
Innovative	In solving urgent problems	Different products
Level of demands	Developing the organization	Concrete
Delegation of responsibility and authority	Learning (continuously and through training)	
Encouragement	Cooperatively with colleagues and management	
Loyalty	*Variation*	
Towards the company	Work rotation	
Towards colleagues	Change in work tasks	
Across organizational boundaries	Flexibility	
Location		
Closeness		
Transportation		
Physical location		
Salary		
Level		
Related to performance		
Ability to support		
Gradually increasing		
The company		
Success		
Size		

(Continued)

TABLE 2.1 (*Continued*)
A Model of the Dimensions and Categories that Influence Work Attractiveness

Work Conditions	Work Content	Work Satisfaction
Security		
Possibilities of advancement		
Benefits		
Relations		
Provides support and empathy		
Team spirit, comradery, cooperation, honesty, directness, and openness		
Social contact		
With customer		
With colleagues		

Source: Data from Åteg et al. (2004), Hedlund (2006).

Mental work refers to cognitive activities. These should be included *in the work task* or *in solving urgent problems*. It is included in *developing the organization*, where the cognitive activities could result in, for example, new ways of performing a task. *Learning* is included in the mental work and should take place both through training and in the daily work. The mental work should be undertaken *cooperatively with colleagues and management*.

The content of the work should also provide *variation*. This encompasses the three qualities of *work rotation*, *change in work tasks*, and *flexibility*. Work rotation means rotation between different work tasks, usually throughout the day. Change in work tasks refers to the continuous enrichment of work; with time, it should come to include more, new, or changed work tasks. Flexibility is the possibility to execute a task in several different ways.

2.2.1.2 Work Satisfaction

Åteg et al. (2004) described that work satisfaction includes the employee feeling *sought after*. This is achieved if the employee feels they have *coveted skills*, are doing *important work*, and in general, if they *feel needed*.

Acknowledgement has an *internal* and an *external* quality. The inner recognition relates to having done a good job, which is based on individual perception (e.g. what it is to do a good job). External recognition represents recognition from others, such as management, colleagues, and customers. The recognition can be in the form of rewards and should be individually based (e.g. if an individual prefers a reward in money, they should get a reward in the form of money; if an individual prefers additional time off, that should be the reward).

Status refers to the qualities *pride*, *success*, and *professional identity*. Pride can come from the employee's own performance or the organization's success. The job should lead to success, which can be within the company, the profession, and so on. The work should also strengthen the employee's professional identity.

The work is stimulating if it provides a *challenge* in such a way that the employee needs to put in effort and use their skill and knowledge to successfully perform it (the employee needs to succeed, but it cannot be too easy). To be stimulating, work also has to contribute to personal *development*; the employee needs to 'get something' out of the job.

The *results* dimension relates to the product of the production or activity. The quality *direct and visible results* refers to a preference of being able to see the results of the work. *Different products* relates to a variation in the products, for example, that each product is unique. The work performed should contribute to a feeling of connection. For example, the employee should know how their work contributes to the quality of the final product. The quality *concrete* stipulates that work should result in something tangible.

2.2.1.3 Work Conditions

For attractive working conditions, Åteg et al. (2004) first mentioned *adequate equipment*. To fulfil this dimension, equipment has to be (perceived as) *modern* and give possibilities to perform one's work with *good quality*, *high productivity*, and with a *healthy workload*.

Working hours are also important for job attractiveness. There should be *predetermined work hours* in the sense that the employees always know when work starts and ends. There should also be an *ability to influence* the working hours regarding aspects such as scope, distribution over the week, and vacation.

The *physical work* environment is a classic topic of study that clearly contributes to work attractiveness. The dimension's qualities deal with *locales* and how these need to be adequate and have a good appearance; *furnishing*, which concerns furniture and decoration; the *air*, which should be good, free from pollutants, and so on; the *sound levels*, which should be low, and *cleanliness*, which stipulate that the workplace should be clean.

Leadership affects working conditions. It deals with *trust* and *communication* between employee and management, and *information* (i.e. that employees know what goes on in the organization). Additionally, management should be *innovative*, that is, wanting to develop the organization as well as its products and processes. Management should also set reasonable *levels of demands*, *delegate responsibility and authority*, and in general, provide *encouragement*.

It is important that the employee feels *loyalty* towards the *company, colleagues*, and across *organizational boundaries*. Loyalty in this case essentially entails being there for either party, making sacrifices, and so on.

Location refers to the geographical placement of the place of work. *Closeness* is important: that the place of work is close to one's home. There should be a variety of *transportation* options for moving between work and home. The *physical location* and its surroundings are also important, such as that the workplace is situated in a nice location. It also includes the whole of the surrounding society.

Salary can be important for attractiveness, where its *level* is important as well as it being *related to performance*, ensures the *ability to support* oneself and that it is *gradually increasing.* Salaries, Åteg et al. (2004) argued, have a tendency to compensate for shortcomings in other areas.

The company itself is important for the attractiveness of the work it encompasses. The attractiveness of the work is likely to be higher if the organization is successful (e.g. economically successful, famous). Its *size* is important; Åteg et al. (2004) suggested that smaller organizations may be more attractive. *Security* contributes to attractiveness. This can be fostered through low turnover rates, a sound economic situation and that there is confidence in the management. It is important that there are *possibilities of advancement* (even if these are not realized) and *benefits* of different kinds (a company car, annual workplace dinners, etc.).

The *relations* at the workplace also play a role. That the relations at the workplace *provide support and empathy* is important. The qualities of *team spirit, camaraderie, cooperation, honesty, directness, and openness* refer to characteristics of the relations. (Note that the original model contains several additional qualities that we have consolidated into two here.)

Finally, *social contact* deals with human contact of different kinds; it can be that the employee has to have the possibility to be in contact with *customers* or with *colleagues*, as a part of the job or during a break. Åteg et al. (2004) noted the difference between social contact and relations. While social contact deals with the presence of other people, relations deal with the interactions of those people.

2.2.2   Objective and Subjective Attractiveness

The models by Hedlund (2007) and Åteg et al. (2004) are mainly concerned with individual and subjective views on industry work. However, we want to bring attention to the distinction between subjective and objective attractiveness. By this we refer primarily to two stipulations. First, that it is possible for an objectively unattractive job to be perceived as attractive (e.g. some idealized jobs) and, conversely, for objectively attractive jobs to be perceived as unattractive (e.g. some hidden jobs). Blauner (1964) illustrated this by distinguishing between subjective and objective alienation in his study of the work of the textile and automobile workers. He found that both the textile and automobile workers were objectively alienated, but only the automobile workers experienced subjective alienation. To Blauner, this was because the norms and values of the societies where the textile mills were located prevented subjective alienation. Similarly, we propose that some workers could be subjectively alienated even though objectively they are not. This was similarly demonstrated by Mayo (1945), in whose study, every change in working conditions (objectively good or bad) increased productivity and motivation.

As for the second stipulation, we suggest that in the long term, objectively attractive jobs may remain subjectively unattractive. Objectively unattractive jobs, on the other hand, will in most cases not remain subjectively attractive. For example, jobs with high demands but where the employee has limited control will result in poor health (Karasek and Theorell 1990). Furthermore, the presence of harmful airborne particles, that is, failure to live up to demands on the physical work environment, also results in negative health effects. According to the model by Åteg et al. (2004), these characteristics should preclude attractive work. Even if a particular individual does not eventually change their perception, other people will change their view (that is, on the whole, the job becomes unattractive). Alternatively, the job will eventually result in poor health. As society becomes aware of this, attractiveness could decrease. We thus propose that some characteristics of attractive work are hierarchical in nature (cf. the hierarchy of needs of Maslow 1987); without the fulfilment of these basic requirements, a job cannot hope to be considered attractive (in the long term). This also implies that factors that *prevent* unattractive work are not the same as factors that *facilitate* attractive work. Therefore, different 'attractiveness factors' are likely to function similar to motivator and hygiene factors (e.g. Herzberg 1968). It is here we also find substance for the argument that creating attractive workplace is not the same as creating safe workplaces.

2.2.3   CONNECTION TO OTHER FIELDS: ERGONOMICS, SUSTAINABILITY, AND WORKPLACE ATTRACTIVENESS

Problems of attractiveness are, in the end, problems of social sustainability. The obvious connection is that attractive work should not cause health problems due to strenuous physical activity, bad air quality, stress, and noise (Johansson et al. 2010a). Essentially, this would degrade the so-called human capital, the prevention of which is often cited as a key component of sustainability in the mining industry (Horberry et al. 2013). Additionally, some mining companies phrase the attractiveness problem as a problem of safety (e.g. LKAB 2017), which other companies, in turn, consider a social sustainability problem (Ranängen and Lindman 2017). Beyond this, workplace attractiveness draws heavily on human factors or ergonomics (cf. Åteg et al. 2004; Hedlund, 2007). In fact, it can be argued that workplace attractiveness is indeed a discipline of ergonomics. Horberry et al. (2013) demonstrated that consideration of human factors or ergonomics in mining contribute to several aspects of sustainability by improving the health and safety of employees, contractors, and the surrounding community; developing effective emergency response procedures; and by facilitating responsible product design and use. Outside of mining, several studies have identified convergent areas between ergonomics and sustainability, as well as how the former can contribute to the latter (cf. Bolis et al. 2014; Haslam and Waterson 2013; Radjiyev et al. 2015). Bolis et al. (2014) specifically connected social sustainability to work-related issues such as worker participation in defining sustainability policies, social inclusion of all types of workers, and promotion of health and safety. Moreover, they identified several studies that noted the possible benefit of introducing ergonomics in sustainability policies to improve the attraction and retention of a qualified workforce. Importantly, Bolis et al. (2014) also identified negative impacts of sustainability policies if focus is limited. They asserted that 'Good planning of sustainability policies cannot exclude the consideration of social aspects' (Bolis et al. 2014, p. 1227), which has implications for social sustainability and attractiveness. While its contribution to social sustainability may appear mostly as indirect, failure to solve the attractiveness problem can lead to unsustainable practices, such as overreliance on a fly-in/fly-out workforce and its associated problem (see Abrahamsson et al. 2014) and development towards one-sided and low qualified jobs (i.e. to lessen the dependence on a *qualified* workforce).

2.3   ATTRACTIVE MINING WORK

The mining industry is not very attractive; with reference to Figure 2.1, the industry occupies the left side of the model. While it is not clear if the mining industry is mostly hidden or unattractive, we consider it necessary for the mining industry to work with both aspects. That is, it needs to improve both its image and work; it is a task of moving the mining industry from its current position in the model to the upper right corner. This gives rise to two questions: '*Can* the mining industry be made attractive?' and, if yes, '*How* can it be made attractive?' We attempt to answer the first question in the following section; we focus on the second question in the rest of the book.

2.3.1   VIEWS ON MINING AND ITS ATTRACTIVENESS

There is some research regarding the mining industry's attractiveness. The majority of that research concludes that mining work is unattractive (especially to young people). Randolph (2011) suggested that the mining industry is unattractive because of remote locations and the less attractive lifestyle such locations offer; the industry image of entailing low-skilled, dirty work in an outdated, boom/bust industry; and a general lack of awareness regarding the opportunities of the industry. Hartman and Mutmansky (2002) claimed that new technology would improve the health and safety of miners by removing danger and thus improve the public image of the industry. Albanese and McGagh (2011) suggested that automation addresses the shorter-term imperative of maintaining a qualified workforce at remote locations as younger generations are reluctant to leave their life in cities where they see their own future. Lee (2011) prescribed training and organizational learning, managing employees' work/life balance, and managing contract workers as components of strategies to solve labour shortages. PricewaterhouseCoopers (2012) contended that the skill shortage in mining is due to entrenched attitudes within the industry and community; entrenched and outdated attitudes towards women preclude solutions to these problems; many rich pools of skills remain untapped; young people are not graduating from relevant educational programs; and many people are reluctant to move to other locations for work. To this end, they recommended that mining companies highlight opportunities in the industry; be accountable for diversity and provide flexibility (in regards to work/life balance) in company culture and roles; address unconscious biases (e.g. regarding gender roles); take care to improve and protect the (positive) image and reputation of the industry; invest in regional infrastructure, provide incentives and remove barriers for relocating; take advantage of new technologies to create more attractive working conditions; and look beyond money to attract and retain employees.

Results from a workshop with around 100 M.Sc. students (Bäckblom 2009) describe the mining industry as dirty, dark, and dangerous; male-dominated; located in remote locations; and as utilizing obsolete technology – with the media playing an important role in fostering this image. However, some students were reported to express positive views of the mining industry, such as it being an interesting, safe, profitable, and high-tech sector. The study also investigated what is required of the mining industry for it to appear as a potential future workplace. The student accounts list the following requirements on mining work: continuous development to improve work conditions and lessen environmental impact; possibility to rotate through several positions of the company to learn all its aspects; responsibility at the consumer level; good leaders; greater gender equality; good communications and broad cultural activities at the location of the mine; possibility for a job for one's domestic partner; and being able to feel proud for working for the company. The students also wanted to see increased automation and remote control, flexibility in terms of working hours and workplace, and improved possibilities to organize their working life.

Zhang and Barclay (2007) found in their study that most mining engineering students perceived the industry as having many job opportunities and being an overall exciting sector to work in. The industry was seen as having relatively safe work environments and technologically advanced. Fewer regarded it as environmentally and socially responsible. The students saw difficulties in balancing career demands with personal relationships and family commitment as unattractive aspects. Their three most important reasons for choosing the mining industry were earning potentials, being able to use their personal strengths, and job security.

Ruiz Martín et al. (2014) investigated the view on the mining industry regarding social impact, environmental impact, infrastructure, industry, employment, housing, government, and communication. They found that the mining industry had positive or neutral relationships with all factors except social impact, towards which the industry showed a negative relationship.

Some research has been more normative. For example, Johansson et al. (2010a) formulated 26 statements that aim specifically to facilitate an attractive mining industry. These statements are listed in Table 2.2 and summarized presently.

Attractive mining has to be safe and should become so through two strategies. The first is the zero-entry mine where human presence in production and development areas are minimized through the use of automation and remote-control technology. The other strategy is to work proactively through systematic work environment management. This entails minimizing risks through systematic and continuous work with risk assessment and other risk management tools. (We return to this topic in Chapter 6.)

The next set of statements is about the physical work environment. While the physical work environment will improve due to, for example, automated or remote-controlled machines that perform the physically heavy jobs, the physical work environment is still important. There still has to be variation in the physical workload; exposure to noise, vibration, chemicals, and radiation has to be as low as possible; the climate has to be comfortable; and lighting has to be sufficient. Locales, machines, and vehicles all have to fit the different needs and limits of humans. (Chapter 3 goes further into this topic.)

To foster a good psychosocial work environment, management has to be supportive of the personnel and the personnel have to appreciate the management. Additionally, there should be cooperation between management and personnel. Instead of planning top-down for individual personnel, people should work in autonomous teams, and a good balance between demands and self-control must be reached. Similarly,

TABLE 2.2

Summary of the 26 Statements for Attractive Mining

Statement	Description
Safety	
Zero entry	Maximum safety is achieved through automation and remote-controlled technology that makes possible zero-entry development and production areas
Systematic work environment management	Proactive and systematic work environment management prevents and controls risks
Physical Work Environment	
Appropriate musculoskeletal workload	An appropriate variation in musculoskeletal load that facilitates physical health is maintained
Minimized noise exposure	The exposure to harmful and disturbing noise is minimized
Minimized vibration exposure	The exposure to harmful and disturbing vibrations is minimized
Minimized chemical exposure	The exposure to harmful chemical exposure (e.g. blasting gases and engine exhaust) is minimized
Appropriate physical climate	The physical climate and heat load are comfortable
Appropriate lighting	Lighting levels provide comfortable light and support the task
Minimized radiation exposure	The exposure to radioactive and electromagnetic radiation is minimized
Workplace and equipment fit for human needs	Locales, machines, equipment, etc. are fit and, based on different human needs and limits, and allow for efficient performance of tasks
Psychosocial Work Environment	
Supportive management	Management is supportive of the employees and are appreciated by them
Management-employee cooperation	There is extensive and efficient cooperation between management and the employees
Work organization based on groups	The work organization is based on autonomous groups and represents the lowest planning level
Appropriate demand-control level	There is an appropriate balance between demands and control, for groups and individuals
Learning	There is learning at the workplace that includes generic, theoretical knowledge to create flexibility in the work system
Holistic understanding	There is a general and holistic understanding of the mining process
Stimulating and challenging work	The work is stimulating and continuously offer new challenges and encounters with new professions
Affirmative action for underrepresented groups	There are actions to affirm underrepresented group and workplace culture that is based on gender equality
Motivating and safety-promoting wage system	The wage system promotes safety and is motivating
Working hours based on social needs	The working hours allow for flexibility based on social needs
Job security	There is job security that is maintained with efficient production
Social Responsibility	
Proudness in company	There is reason for employees to feel proud to work for the company
Living mine site	The mine site is connected to a living society with broad cultural activities

(Continued)

TABLE 2.2 (*Continued*)
Summary of the 26 Statements for Attractive Mining

Statement	Description
Minimized FIFO	The utilization of FIFO workers and contractors is minimized
Equal rights for contractors	Contractors and the company's employees have the same rights and obligations
Minimized environmental impact	The environmental impact is minimized

Source: Based on Johansson et al. (2010a).

Hutchings et al. (2011) found that improved employee engagement is seen as an important change for the Australian resources sector to appear more attractive. Work in mining should mean continuous learning: both theoretical knowledge and an understanding of mining and its operations in a holistic perspective. Mining work should offer new challenges and encounters with new professions. Hutchings et al. (2011) also found that the Australian resources sector sees this as a need. In attractive mining workplaces, there should also be gender equality (as well as equality in general), and actions that promote equality should be undertaken. A job in the mining industry should also entail job security grounded in efficient production. Wages should discourage risk taking and work hours should be flexible enough to respond to social requirements. (Chapter 5 expands on these issues.)

The social responsibility of the company should include making sure employees feel proud to work for the company. The mine site should be connected to a living society with broad cultural activities. A consequence of this is avoidance of fly-in/fly-out (FIFO) personnel and contractors where possible. (Chapter 5 also discusses FIFO operations and contractors.) Where utilized, FIFO personnel and contractors should have the same rights and obligations as the company's employees. Mining companies have to minimize its environmental impact.

2.3.2 CAN THE MINING INDUSTRY BE ATTRACTIVE?

The discussion so far has focused on what the mining industry should do to become attractive, but has not covered the possibility of the industry to enact these changes; the discussion here aims to show the potentials for the mining industry to be attractive. We depart from the model by Åteg et al. (2004) and discuss the mining industry's current state as well as its potential future state. Our focus is on operator jobs, but the discussion extends to and is relevant for other jobs as well. This includes professions such as repair or maintenance workers but also white-collar professions. For the latter, the account from the different students above should indicate the applicability of the discussion.

2.3.2.1 Work Content of Mining Jobs

The *working speed* in the mining industry tends to be intense. In part due to high investment costs, machines often have to be manned at all times to ensure a return on the investment. Additionally, mills need to be continuously fed, contributing to similar tendencies. In some cases (e.g. in mass mining), operators are transported to and from load-haul-dump trucks (LHDs) at shift changes and breaks to ensure that the LHD is run as continuously as possible; the truck cannot afford to stop. In such scenarios, intensity is high with few calm periods, even though longer breaks are possible. However, the situation in mining is more variable than that: other mining methods and activities can offer more opportunities for breaks and calm periods. Regardless, mining does have the potential for providing periods of intensity and calmness as well as pauses. Loading operations provide a good example. Some mines have developed their loading operations to be remote controlled and semi-automated. These mines have systems that automatically drive the LHD between the front and the chute. The operator is only involved when actually loading and dumping. This allows the operator to operate several LHDs simultaneously. At the same time, this could allow for more calm periods and shorter breaks as the operator is not physically 'tied' to the machine.

Even though mechanization has removed a lot of *physical activities* in mining, they are still common. The model by Åteg et al. (2004) stipulates that this activity should be *healthy*, but much evidence suggests that it is not. For example, the Swedish Work Environment Authority (2016) listed the mining industry as still significantly affected by occupational diseases caused by physical workloads. Even if this figure is variable – a subsequent report (Swedish Work Environment Authority 2017) listed the mining industry as having relatively few occurrences of occupational diseases caused by physical workloads – McPhee (2004, p. 297) stated that 'Heavy physical workloads and stresses are still areas of concerns, but are likely to be intermittent rather than constant'. Elgstrand and Vingård (2013) listed the prevention of musculoskeletal disorders as one of the current needs of the global mining industry. Thus, while there is definitely physical activity in mining, it is not always healthy. Furthermore, a common motivator for increased mechanization and automation in the mining industry is the removal of unhealthy physical activity. This is positive but tends to remove *all* physical activity. This can be negative from an attractiveness and health perspective (e.g. due to increased sedentary work). Moreover, mechanization may not effectively prevent issues relating to physical workloads. For example, problems stemming from whole-body vibrations become more prevalent with mechanization (McPhee 2004). Work from control rooms (the result of the higher levels of mechanization and automation) tends to involve limited physical activity. Even so, preventive maintenance is common in the mining industry. This provides opportunities for physical activity in mostly mechanized operations; as long as maintenance can be performed under good ergonomic conditions, it could be considered healthy. Highly automated systems are even more dependent on preventive maintenance (Lynas and Horberry 2011). Even in this type of mine, then, there should be opportunities for healthy physical activities realized through practices such as work rotation.

Spatial movement is related to physical activity. This quality depends on which activity and type of mining is conducted. Currently, chargers, for example, can experience a great deal of spatial movement during a workday. A loader in sub-level caving mines, in contrast, may be confined to their LHD for the entire workday. The increased control-room work of the future may decrease the natural spatial movement, but like the other physical activity, can increase with work rotation.

Familiarity is not dependent on the specific work task but rather on the organization. The sizes and distances in mining can make communication of information relating to what to do and what to expect more difficult, but does not prevent it. Additionally, the digitalization of the mine makes communication of this type easier: relevant information can be accessed directly through, for example, a smartphone. Work from control rooms in general also makes this type of communication easier. This is also true for *freedom of action*. That is, it depends more upon the organization rather than specific work tasks. Here too, digitalization and extended work from control rooms can provide more opportunities for controlling and organizing one's own and others' work.

The use of handheld tools and machines is increasingly rare in mining (Järvholm 2013). In modern mining, *practical work*, as described by Åteg et al. (2004), exists mostly in charging and maintenance (Abrahamsson and Johansson 2006). With increasing automation, it may only exist in maintenance, which, on the other hand, will be extended. Again, a system of work rotation would be a way of ensuring practical work for employees. The practical work should include opportunities for *creativity*. Currently, retrofitting and a variety of ad hoc solutions are common in mining, which could offer opportunities for creativity. Highly automated systems might be less accepting of such practices, but Horberry and Burgess-Limerick (2015) argued that many mining workplaces are so complex that it is impossible to determine all interaction within the sociotechnical system in advance; all systems will continue to need some sort of redesign and retrofit to function optimally. This opens up for creativity in future operations as well.

Mental work is present, in part, in current mining activities (perhaps mostly in solving urgent problems and in retrofitting work). In controls rooms, there are many opportunities for mental work. Some mining companies also involve their employees in developing the organization (Lööw 2015). Continuous learning and cooperation with colleagues and management are organizational issues and do not necessarily depend on the specific work task, though distances can complicate gathering the workforce for such cooperation or activities (Haugen 2013). Here as well, digitalization and extended control room work can facilitate such activities. The future mining industry offers many possibilities in regard to mental work. Abrahamsson et al. (2009) stipulated that as automation increases and operators are moved to production centres, their jobs will change in character: towards service work and tasks that require a variety of skills (including theoretical skills). Abrahamsson et al. (2009, p. 309) put this in terms of there being 'an emerging and, in many aspects, already evident knowledge transformation – from the old and obsolete physical and tacit knowledge and skills ... to something new which can be described as abstract knowledge'. Such knowledge work is bound to include continuous learning and will probably also require constant on-the-job training as the advanced systems evolve.

Variation is also a dimension that to a large extent is determined by work organization, and we have already noted *work rotation* for allowing for other qualities. The dimension depends on the width in variation, of which there is plenty in mining. As mining continuously becomes more automated and mechanized, a change in work tasks will follow. Currently, there is *flexibility* in the different tasks of mining. Flexibility may be problematic in the future as the new systems might demand very standard and algorithmic operations, but again, with work rotation there is a possibility of ensuring that operators perform tasks that are both rigid and flexible.

2.3.2.2 Work Satisfaction in Mining Jobs

In the mining industry, the dimension of *sought after* is for most qualities satisfied or has the potential to be: the background for this book is that the mining industry needs people, especially skilled people. This does not directly translate into employees *feeling* needed, so there may be a need to work with facilitating this feeling. Fostering perceptions of mining as important work is not a work-task specific challenge but is dealt with at the company or even societal level.

Acknowledgement is a challenge on the organizational and individual level. *Internal recognition* depends on the individual's perceptions and cannot really be actively influence other than making sure the work task has space for personal 'interpretations' (e.g. regarding execution). *External recognition* can be facilitated through further digitalization or through a consolidation of the workforce (i.e. locating it in control rooms or production centres) so that there is a more direct contact between the operator, other operators, management, and the customer.

Status, too, relies little on the specifics of the task and depends instead on activities at company and societal levels. The transformation of mining work may challenge perceptions of the professional identity. This could be negative for current employees in terms of attractiveness but may very well act positively for a new generation of employees.

Currently, it may well be that mining is *stimulating*; the perception of it being *challenging*, *developing*, and *interesting* is dependent on the individual. Some individuals no doubt consider mining work to be all this. Future mining should also have ample opportunity to be stimulating, though there might be a mismatch between the current and future workforce.

Regarding *results*, the product is temporarily visible for only some employees; a charger only really sees the front while a loader sees the ore; the product is not final until leaving the mill, several operational steps later. This can make it hard to glean a feeling of connection. Furthermore, the product is one and the same for a given task. Work in control rooms, where all operations have a closer physical proximity, may help in this. The same is true for increased digitalization that offers the opportunity to visualize each operator and operation's contribution to the final product (e.g. how certain action will affect to quality of the outcome).

2.3.2.3 Work Conditions in Mining

With regards to *adequate equipment*, mining today covers the whole spectrum of manual, handheld tools at one end, to large, highly advanced, automated machines on the other. Even so, Horberry et al. (2011) showed that much mining equipment is lacking in their design and consideration of human factors. This impacts quality, productivity, and the possibility for healthy workloads. Of course, some equipment is fully adequate. The future of the mining industry holds much potential in this aspect: designing mining equipment and workplaces with full attention to the human will ensure that all qualities for equipment are satisfied, regardless of whether work takes place underground in machines or in production centres. Though in terms in attractiveness, there is an additional challenge: the equipment must also be adequate for the new workforce.

Working hours is a complex topic for mining where the use of extended work-days is common. For example, Dembe et al. (2005) showed several negative effects associated with long work hours, such as higher injury and hazard rates. Cliff and Horberry (2008), on the other hand, did not find any strong associations between hours of work and the number of incidents and injuries. It is complicated further by the use of a FIFO workforce. The model by Åteg et al. (2004) stipulates that employees should be able to control their working hours. However, such control should not compromise the safety or health of the employee. But mining operations usually require round-the-clock operation. This places restrictions on working hours. In the end, these questions tend to depend on factors such as the location of the mine (i.e. the distance to societies, which influences the need for a FIFO workforce) and organizational decisions. Some predictions for future mining sees remote operation centres located in major populated areas. This could greatly influence this topic.

The *physical environment* is a traditional topic in mining. *Air quality* and *noise levels* remain problematic (Elgstrand and Vingård 2013), and maintaining *cleanliness* is understandably challenging. Air quality and noise levels have improved though, especially in cases where machines with isolated cabins have been introduced or work has moved to control rooms. While cabins tend to be exposed to dirt, control rooms can maintain cleaner environments. Future mining can satisfy these qualities through further remote control and the increased introduction of machines with isolated cabins. The dimension, as described by Åteg et al. (2004), also deals with the characteristics of the facilities and locales of the workplace, including changing rooms and break rooms. While work continues underground, the operational environment is difficult to change to fully satisfy these requirements. Control rooms, on the other hand, offer that opportunity.

Leadership is important and is often recognized as such in the mining industry. It is unrelated to the specific work tasks and is dependent on the organization. Like other issues that in the end relate to communication, the distances involved in mining can preclude efficient contact. Again, the consolidation of the workforce to control rooms and the further digitalization of the mine can help facilitate communication and, thus, the qualities related to leadership. Qualities relating to *loyalty* are similarly dealt with on an organizational and a company level.

A mining company is essentially unable to control the *location* of a mine. However, many mining companies provide transports to and from the mine (be it within the local community the mine is located at or to and from the remote locations of the FIFO mines) and can thus influence this topic.

Salaries in mining are generally good. Even at lower levels in mining organizations, they are significantly above average. Åteg et al. (2004) reported that salaries should be related to performance to contribute to attractiveness, but this may be complicated in mining. Johansson et al. (2010b) found mostly negative effects of piece-rates on health and safety. Additionally, Kronlund et al. (1973) described the effects of removing a piece-rate-based wage system in a Swedish mine: decreased severe accidents by 95 per cent and normal accidents by 70 per cent (although increased minor accidents by 45 per cent).

The qualities relating to the dimension of *the company* is only marginally affected by the design of mining work. But the breath of activities in mining should allow for several different possibilities of advancement. The *size* of the company is hard to affect as it depends on the orebody, the mining method, and so on. However, it may be relevant for the sizing of work teams, for example. Additionally, it may be a relevant question in regard to contractors. The rest of the qualities depend on company-level decisions.

The same is true for the dimension of *relations*. The existence of teams in mining may facilitate the relevant qualities, though work often also binds an operator to a single machine. It is possible that the qualities can be better fostered in control rooms or with the help of extended communication systems. The *social contact* dimension is also similar to this: the pursuit of these qualities should be easier when facilitated by control rooms and further digitalization.

2.4   SUMMARY AND CONCLUSIONS

While the subject of workplace attractiveness itself is quite new, it has clear connections to other rich research traditions and subjects. From these other fields, it is clear that technical questions, and questions of providing work satisfaction and creating rich work, are questions of a social nature: it is the consideration of the operator, the human, and designing on their terms, in combination with technical and organizational aspects, that will facilitate a solution to the problems of concern in this book (cf. Bohgard et al. 2009).

Almost every aspect of work influences its attractiveness. Thus, it has to be approached holistically. In this chapter, we discussed the theoretical framework from which we depart, while the rest of the book will delve into specific issues and ways of working with them. Essentially, we can more or less say what constitutes attractive work, but so far, there is less written regarding how to work with these issues. For example, PricewaterhouseCoopers (2012) argued that mining companies should be accountable for diversity and provide flexibility, but the 'how' of it needs elaboration. For these and similar issues, we hope to provide important insights. Additionally, we refer frequently to job rotation to facilitate attractive workplaces. This may be diffuse. What we mean is that jobs can be designed in such a way that they balance out unattractive aspects with attractive aspects, for examples, through the use of work rotation. It is our hope that the following chapter will help illuminate this notion.

3 Health and Safety in Modern Mining

3.1 ON DEFINING AND SPECIFYING HEALTH AND SAFETY IN MINING

The subject of health and safety in mining is a well-researched topic and has been, in some surveys, identified as the most important challenge for mining companies (Ventyx 2012). There are several important publications within this area, and so, we don't intend to give another traditional account of health and safety in mining. Instead, our intention is to provide an alternative and sometimes overlooked perspective. We take our starting point, the contribution of health and safety to the notion of attractive workplaces, and attempt to demonstrate how social and organizational aspects are key issues in this endeavour. That is not to say that we are arguing against the traditional view on health and safety, but rather we expand this approach by including important additional perspectives.

Saleh and Cummings (2011) argued that it is common in the mining industry to identify issues of health and safety separately. They stated that 'This separate classification does not necessarily reflect mutually independent hazards, but it helps recognize a difference in the time scales of effect of the hazard sources' (Saleh and Cummings 2011, p. 768). This is summarized in Figure 3.1; Saleh and Cummings (2011) exemplified this view with explosions or mine collapses that will immediately result in traumatic injuries and fatalities; on the other hand, prolonged exposure to harmful particles (e.g. coal or silica dust) can result in fatalities as well but that takes years to happen.

The distinction by Saleh and Cummings (2011) allows for health and safety risks to have the same source (i.e. stem from the same hazard). While the prevention of each hazard or risk tends to vary depending on its nature, the same principal strategies for control and prevention can be applied. Chapter 6 deals with the specifics of these strategies, but they are reviewed briefly here to introduce the terminology. The strategy is commonly known as the hierarchy of control and exists in several different conceptions. Haddon (1973) drew attention to ten countermeasure strategies; they are summarized here as three levels of control:

1. In the first level, risks or hazards are completely eliminated. An example of this is eliminating the risk of falling by performing work at the ground level.
2. In the second level, risks or hazards are minimized or reduced by decreasing the likelihood and/or the consequence of the risk. This can be done through substitution, isolation, or engineering controls.

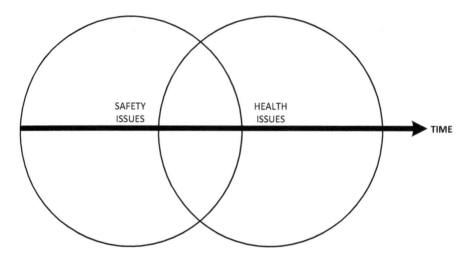

FIGURE 3.1 A graphical representation of the difference between safety and health issues. (Based on Saleh and Cummings 2011.)

3. In the third level, personal protective equipment (PPE) or administrative controls are used. These also reduce the likelihood and consequence of risks but are aimed at the person.

Another note on terminology and definitions concerns the concepts of 'risk', 'accident', 'hazard', and 'safety'. Health and safety research (including accident research) has a long history. As was noted in the 1960s by Haddon (1963), the terminology, definitions, and even underlying assumptions can vary significantly between authors, subjects, and so on. 'Risk', which in everyday usage tends to refer to probability but usually has precise definitions, is perhaps the best example of this. To avoid confusion, what follows is a short review of the terminology utilized in this chapter:

- *Risk*. Risk is the product of the probability of an accident happening and the consequence if that accident were to happen, sometimes expressed as $R = \sum p_i\, C_i$ (where p_i is the probability of occurrence and C_i is a measure of the consequence; Harms-Ringdahl 2013). (While we avoid this mathematical definition when possible, risk should still be viewed as the combination of probability and consequence.)
- *Accident*. 'An accident is an event that causes unintentional damage or injury' (Harms-Ringdahl 2013, p. 13; cf. Saleh and Cummings 2011). In safety statistics, 'accident' can also refer to the severity of the injury (e.g. an accident is an event that causes an injury and at least 5 days of absence from work); in other cases, severity is used to distinguish between different types of accident (i.e. the consequence of the accident).
- *Hazard*. A hazard is a source of risk (Harms-Ringdahl 2013); without a hazard there can be no risk.

- *Safety.* Safety is not simply the absence of accidents. Beus et al. (2016, p. 354) argued that 'Whereas accidents clearly indicate the absence of safety, a lack of accidents cannot necessarily be used to infer the presence of safety' and thus defined safety as 'an attribute of work systems reflecting the ... likelihood of physical harm ... to persons, property, or the environment during the performance of work' (Beus et al. 2016, p. 353). Thus, high safety means a low likelihood of physical harm, and the other way around.

3.2   HEALTH HAZARDS AND RISKS IN MINING

When it comes to work-related health problems, the mining industry is overrepresented. The European Commission (2010) found that in 2007 the mining industry was the economic sector with most work-related health problems in the European Union. The same study also showed that work-related health problems in mining had gone up, and while health problems had increased in all sectors, they had significantly risen more in mining than in most other sectors. Though the situation in mining varies between countries and companies, even when doing comparatively well, mining is often worse than other sectors (i.e. a mining company might have a good health record by mining industry standards but can still be above average in national statistics).

The most common health problems in mining are respiratory diseases, noise-induced hearing loss, and musculoskeletal disorders (Elgstrand and Vingård 2013). These are consequences of physical, chemical, and ergonomic hazards. Physical and chemical hazards are primarily generated by the environment in which mining takes place. Ergonomic hazards are more dependent on the design of, for example, machines and work tasks. In a longer term perspective, some important changes have occurred in terms of frequency and relative occurrence of work-related diseases. Table 3.1 compares the difference in the most prominent factors of work-related diseases between the 1980s and 2010s in the Swedish mining industry and shows a change in causes of occupational diseases: from chemical (no biological factors for the mining industry were recorded in the 1980s) and physical (i.e. noise and vibration) factors in the 1980s, to ergonomic factors (i.e. physical ergonomic factors) in the 2010s. This is consistent with developments in the industry: exposure to noise and chemical factors decreased as mining became more mechanized and automated and work began to a greater extent be performed from isolated cabins. But during this period, ergonomic issues increased. As is argued below, this may be explained with reference to the fact that many mining machines still lack in several ergonomic aspects. On the other hand, Table 3.1 also presents the aggregated frequency of the causes of health issues in Swedish mining during the two periods. These data show that all but one of the relative frequencies have decreased (ergonomic factors are only relatively more prominent). The only marginally visible increase is that of factors relating to organizational and social factors (and these figures are uncertain due to the low numbers reported).

Workplaces of the Swedish mining industry are healthier now. But there is still work to done, especially as Swedish mining companies are likely among the "healthier" mining operations in the world (meaning many other are worse off).

TABLE 3.1

The Aggregated Number, Frequency, and Distribution of Causes of Work-Related Diseases in Mining in Sweden

	Ergonomic Factors[a]	Chemical/ Biological Factors[b]	Noise/ Vibrations	Organizational/ Social Factors	Other[d]	Sum
			1980–1984			
Count	236	200	307	3	28	774
Frequency	3.8	3.3	5.0	–	0.5	12.6
Proportion	30%	26%	40%	–	4%	
			2010–2014			
Count	82	14	54	4	13	167
Frequency	2.0	0.3	1.3	0.1	0.3	4.1
Proportion	49%	8%	32%	2%	8%	

Source: Data from National Board of Occupational Safety and Health (1983, 1984, 1985, 1986, 1987), Statistics Sweden (1982, 1983, 1984, 1985, 1986), Swedish Work Environment Authority (2015, 2014, 2013, 2012, 2011).

Notes:

[a] The official translation is 'ergonomic factors' but is close to physical ergonomics.

[b] Earlier publications separated these categories (but no biological factors were reported in the first period).

[c] Earlier publication separated these categories into 'noise' and 'vibration' while later publications merged them; here, the categories are merged.

[d] Earlier publications listed 'other physical factors', but they have been merged into this category.

Additionally, health problems seem to be migrating from one area to another. To be attractive, it is important that mining workplaces are healthy. And it may appear that creating healthy mining workplaces primarily affects internal attractiveness. However, external attractiveness is also affected by health aspects of the workplace. Abrahamsson et al. (2015) discussed the exporting effects: what happens in the workplace also affects the surrounding community – the effects are exported. This is perhaps most clear with work-related health problems. Such problems, engendered at the workplace, spill over into the surrounding society by producing a requirement for health services or decreasing the quality of life. Through its association with such negative effects, the external attractiveness of mining companies can decrease. Finally, there is also a connection between attractiveness and the prevention of occupational diseases. Poor work environmental conditions increase the turnover rate of personnel, which can effectively hide issues related to occupational diseases (Gardell 1976) because many of these diseases take some time to manifest themselves (if reactive methods are relied on to signal the existence of health hazards, this might be an especially pressing issue; see Chapter 6).

In what follows, we present more details regarding the health issues of the mining industry and its relation to ergonomics and attractiveness (and to some extent, also safety). Yet, our intention is not to present an extensive review; there is already an extensive literature that covers the subject. Rather, the topics we raise below are what we consider as good examples of the nature of the problems that the mining industry has to solve; they work well in outlining requirements regarding their solution.

3.2.1 Dust

In mining, air quality largely regards dust and its resultant respiratory diseases. Many different respiratory diseases occur in mining: Elgstrand and Vingård (2013) and Donoghue (2004) listed coal workers' pneumoconiosis, silicosis, asbestosis, emphysema, and chronic bronchitis as common. These are caused by exposure to the airborne dust found in mines. The dust in mines often includes silica, coal, radon, and diesel, which cause these diseases. Many of these particles also increase the risk of lung cancer. Dust is also problematic because it limits visibility. This can increase the likelihood of accidents, decrease work performance, and so on. (Air quality also includes climate, but the points we want to make are best illustrated with reference to dust and similar particles.)

Air quality is part of the work attractiveness model put forward by Åteg et al. (2004). Subsequent research has revealed that, in the Swedish context, air quality is, or should be, one of the most prioritized areas of workplace attractiveness (Hedlund et al. 2010). That is, it is one of the areas that contributes significantly to workplace attractiveness, but that is currently only to a low extent fulfilled. Furthermore, Carayon and Smith (2000, p. 651) noted that 'air quality … [has] been shown to affect energy expenditure, heat exchange, stress responses, and sensory disruption which make it more difficult to carry out tasks and increase the level of worker stress and emotional irritation'. Additionally, ventilation costs in mining are significant (Atlas Copco 2014); to the extent that dust control becomes less reliant on ventilation, there can be both economic and environmental savings. In turn, this can increase attractiveness (Jäderblom 2017).

Respiratory diseases are, regardless of disease, effect, or type of dust, best controlled by controlling the materials that cause these diseases. Dust in mining is generated when mining takes place (e.g. when the rock is fragmented). Thus, it is seldom possible to eliminate the source of the risk because this would mean not fragmenting the rock. On the other hand, the risk can often be reduced or isolated. To achieve this, technical solutions are often used.

3.2.1.1 Current Controls and Solutions

Ventilation reduces the amount of dust that personnel are exposed to, and thus, reduces risks. Kissel (2003) listed two types of ventilation: dilution and displacement. *Dilution ventilation* works through diluting the dusty air by providing clean air. This means that the air contains a lower percentage of harmful particles. With *displacement ventilation*, an airflow is used to 'trap' the dust at the source and keep it away from personnel; the intention is to keep the dust downwind of workers. Kissel (2003) also mentioned water spraying for dust control. This can prevent dust from getting into the air after it is generated (wetting) as well as capturing dust that is already airborne (airborne capture). Additionally, dust collectors that function like vacuums cleaners can be used, for example, in cabins or at cutting heads of roadheader.

The less dust that is generated in the first place, the better. Accordingly, dust generation should always be reduced wherever possible. Strategies to accomplish this include using deeper cuts in long-wall operations, using water-injection drilling, and preventing personnel from entering the production area after blasting to let the dust settle or be removed by the ventilation (Kissel 2003). Planning is sometimes used to control exposure. For example, in mines where radon exposure is a problem, mining companies tend to rotate personnel to keep their cumulative exposure below threshold values. According to Kissel (2003), dust can also be generated by moving equipment, as trucks moving loads on poor roads, and the roads themselves, produce dust. In both cases, it is important to have well-maintained roads of high quality.

Some dust can be removed through substitution. An example is using electricity-powered machines instead of diesel-powered mining machines, which reduces exposure to diesel fumes. Low emissions diesel fuels and engines with a high European emission standard (e.g., Euro VI) also reduce diesel fumes.

Finally, automation and remote control are effective solutions for lessening operator exposure to dust. Moving just a short distance from dust sources can lessen exposure (Kissel 2003). Control rooms are much less dusty than production areas. Mechanization, with operation in an isolated cabin, also decreases exposure. However, dust can still enter through, for example, dusty or dirty clothing, which makes good housekeeping practices and keeping doors closed during operations important issues (Horberry et al. 2011).

3.2.1.2 Designing for Dust Control

As noted, it is the mining environment that gives rise to the dust problem. While the operator can be moved away from this environment (to a control room, for example, or further away from the source of dust), it is more seldom that the environment itself can be changed. Dust can most readily be controlled through technical solutions: ventilation, wetting, use of battery power, and so on. Some task-related aspects can also decrease dust generation and exposure. Examples include speed limits and organizational aspects such as production scheduling and routines for preventing dust entering clean areas. However, the effectiveness of task- and organization-related solutions depends on social factors. Among these are organizational climate and the general level of stress. In the case of stress, that can in part be caused by the air quality and production planning. In turn, this can lead to routines, put in place to prevent dust generation, being ignored. Another complexity is the issue that solutions and problems tend to 'feed into' each other: a solution to one problem may in turn create other problems. Wetting as a method for dust control, for example, means spraying dusty material with water, and generally, the more water that is used, then the more the dust is reduced. However, too much water can be a problem: not only can it have a negative effect on operations due to issues relating to quality and handling material, but also on safety as it creates slippery surfaces. Another example is scheduling production to reduce exposure to, or the generation of, dust. Such interventions may result in stressful schedules (e.g. reducing the time in which a task has to be completed) that can decrease the effectiveness of other measures.

Moving work to control rooms or isolated cabins is an effective way of greatly reducing the dust exposure. Nonetheless, this is not always or fully possible; even in highly automated mining, it is important to remain mindful of different dust control strategies. Additionally, as noted, dust can still find its way into dust-free environments, and this requires, for example, well-planned routines. In these cases, there is also the problem related to the fact that those people who generate dust are not those people that are exposed to it. For example, limiting speed reduces dust generation. But when a vehicle is operated from an isolated cabin, there may be less incentive for the operator to keep dust generation at a low level by lowering speeds; the situation is different, however, for people working outside the cabin.

In addition to being a prerequisite for attractive workplaces, solving issues related to air quality can have positive consequences for productivity and economic factors. Lööw et al. (2017) found that some mining companies invest in automation and remote control to be able to enter production areas more quickly after blasting. Similarly, if the air quality is improved (e.g. dust generation is reduced), personnel can remain in production areas longer. Performance is also increased solely on account of enhanced air quality; the environment generates fewer stressors, depletes fewer resources, and leaves more resources to be 'spent' on performance (Carayon and Smith 2000). This is in addition to costs saved due to reduced ventilation demands.

The point being made here is that it is not enough to rely only on technical solutions, and that there are many advantages in improving air quality through a holistic approach. A holistic approach entails investigating several different methods of reducing dust, considering both 'hard' and 'soft' factors, while being aware that each solution to a problem may generate problems in other areas. Working systematically and involving all stakeholders is the key to a holistic approach; we cover this further in Chapters 6 and 7.

3.2.2 Noise

Noise is common in mining, and most activities produce some sort of noise: drilling, blasting, handling materials, ventilation, crushing, conveying, ore processing, and so on. The (prolonged) exposure to noise has short- and long-term effects. Bohgard et al. (2008) listed three negative effects of noise other than permanent and temporary hearing reduction and loss. The first effect is the subjective disturbance caused by noise. This disturbance can result in feelings of discomfort, disturbed sleep, and reduced performance. The second effect is the physiological reaction to noise, which can be divided into three categories aside from causing stress. The first reaction is towards (unexpected) changes in the sound environment to which attention is diverted. This reaction diminishes as the change reoccurs. The second reaction is towards high sound levels. This reaction is similar to the first reaction, but no diminishing effect occurs. The third reaction is persistent and includes increased blood pressure, increased muscle tension, and the release of stress hormones. Long-time exposure can make some of these reactions permanent. The third effect of noise is the masking of conversations and warning signals. This masking effect reduces the possibility for information exchange and increases the risk of accidents.

As with air quality, noise levels are important for attractiveness (Åteg et al. 2004). Carayon and Smith (2000) claimed that noise is the most well-known environmental stressor in relation to the effect it has on the outcome of a work system. Even though knowledge about the problems is far from new, and despite the existence of many regulations, noise remains a significant problem within mining (McBride 2004). Reeves et al. (2009, p. 5) argued that despite the mining industry's recognition of its importance, the noise problem prevails:

> … because of the relatively small market for mining equipment, manufacturers have limited incentives to develop less noisy machinery or more innovative noise controls. Also, the specialized equipment designs imposed by the sometimes-hostile mining environment has limited the transfer of noise control technologies from other industries.

Compared to air quality, noise cannot be removed or controlled through, for example, mine infrastructure (e.g., ventilation). In this sense, mining companies can have less control over this issue; coupled with the many sources of noise, it is difficult to control all of the sources.

3.2.2.1 Current Control and Solutions

Dealing with noise can be reduced to three aspects: source, path, and receiver. The idea is that the *source* of the noise can be eliminated, reduced, or isolated; the *path* of the sound can be altered, thus reducing noise; and finally, the *receiver* (in this case, usually the personnel) can be protected and isolated (Reeves et al. 2009). Eliminating the noise completely is the most preferable option but seldom possible because the source of the noise has to be removed (which is blasting, shearing, etc.). This has made controls through engineering (so-called engineering controls) common.

Barriers and sound-absorbing materials are commonly used controls. These should be placed as close as possible to the source of the noise and made as tall and wide as possible. This way, the barriers can better interrupt the path of noise from the source to the receiver. Their design (e.g. shape and materials) have significant bearing on their effectiveness. Using enclosed cabs is an effective engineering control that combines barriers and sound-absorbing materials. This means that efforts to procure well-designed machines can have significant effects on reducing the problem of noise in the workplace. Retrofitting machines is possible, and common, but often more expensive and less effective (Reeves et al. 2009).

Reeves et al. (2009) also listed isolation and reduction controls; for example, sound from an engine can be reduced by installing (better) mufflers, and noisy machines can be spatially or physically isolated. Remote control can reduce exposure to noise or completely isolate personnel from it.

Hearing protection is commonly used in mining but should be considered as a last effort because they are often misused or not used at all. It can also be difficult to ensure they fit the personnel, and all are not very efficient. When using PPE, it is also more difficult to communicate and warnings may not be heard (McBride 2004).

Administrative controls are also available. This involves reducing the exposure of personnel to noise by decreasing the amount of time they spend in noisy environments or by controlling the source of noise (e.g. a certain machine is not allowed to be used when there are many personnel in the vicinity).

3.2.2.2 Designing for Noise Control

The noise problem in mining stems largely from the utilized machinery and its (or other tool's) interaction with the rock. While it is possible to lower the sound levels of most machines (e.g. by using battery or electric power instead of diesel), this alone is not enough to solve the noise problem. Battery-powered loaders have sound levels as low as 80 dB (Jäderblom 2017), which is significantly lower than the sound levels of a diesel-powered machine. Nonetheless, this level is above acceptable sound levels. While acceptable noise exposure level varies between countries, a time-weighted average (TWA) is normally used. The permissible noise-exposure time limits established by the Mine Safety and Health Administration stipulate that the exposure duration for a TWA of 90 dB is 8 h, 6 h for 92 dB, 4 for 95 dB, and so on. It also stipulates that at the TWA of 85 dB, hearing protection has to be offered to the worker (Reeves et al. 2009). In Sweden, regulations list a lower and upper threshold value. The lower value is at a TWA of 80 dB and requires the employer to offer employees hearing protection. The upper value of TWA of 85 dB requires the employee to wear hearing protection. Regulations of the European Union for TWA levels are similar, stipulating that information and hearing protection has to be provided at 80 dB and TWA of 85 dB cannot be exceeded (Bohgard et al. 2009).

Furthermore, the existence of regulations in relation to noise has not prevented the extensive noise problem in mining (this is apparent in statistics that depict an over-representation of occupational diseases caused by noise in mining; e.g. Swedish Work Environment Authority 2017; Elgstrand and Vingård, eds. 2013). This suggests two things: that even when below the threshold value, noise has negative effects; or that hearing protection is insufficient as a solution. McBride (2004, p. 292) stated:

> Hearing protective devices … are often used, and misused, in the mining environment. … compliance issues arise because [they] are not particularly comfortable and interfere with communication, so if noise is intermittent – always a problem in mining – they are unlikely to be worn or may frequently be taken off.

Even at low levels, 'noise can reduce comfort and increase annoyance [and] so, potentially, influencing work performance and well-being through, for example, lowering of [an] … operator's concentration' (Horberry et al. 2011, p. 82), apart from contributing to health problems.

The complexity of the issue is that it appears unlikely that the noise level of mining machines will be reduced to a level that they do not warrant any sort of hearing protection or noise control; even if mining companies would fully utilize electric machines, operators in the vicinity would still need isolation or protection from the noise to avoid negative health impacts, decreased performance, and so on. This means that either all operators must perform most of their tasks from an isolated cabin (or control room), and only perform tasks outside of their protected environment when machines are shut down, or for short periods of time and while wearing hearing protection; or that noise-reducing measures (mufflers, barriers, etc.) and the use of personal hearing protection is extended. There is always the risk that hearing protection is not used. In combination with decreased auditory perception and hindered communication, the risk of accidents can increase. Extending the use of different engineering controls can also be difficult as many mining machines are mobile; while barriers and similar measures can technically be installed on machines, it may not always be practically feasible to do so because of lack of space, operating restrictions, and so on. Reeves et al. (2009) demonstrated the limited gains in most control strategies. However, their studies also found that a fully enclosed environmental cab can provide 20 dB of noise reduction, sometimes more and that even a partly (three-fourths) enclosed cab can provide substantial noise reduction.

When designing mining work, ensuring appropriate noise levels is an issue of taking a holistic approach (as when designing for air quality). This sub-section attempts to illustrate that no one solution is sure fire, but that a combination is likely to be required. Work rotation may appear as a suitable solution as it can decrease daily average noise exposure and function to make a job more attractive. But noise exposure has immediate effects as well. That is, while negative long-term effects can be decreased, short-term effects, such as increased stress and decreased concentration, would still affect the task in question; noise reduction measures need to be taken even in cases where exposure is short or low. This assumes that technical measures are used in combination with, for example, organizational measures.

3.2.3   Physical Ergonomics

There are many ergonomic risks associated with mining. At the same time, however, due to changes in mining work practices, physical ergonomics are now of a different nature than they have traditionally been. McPhee (2004, p. 298) pointed out that 'Physically heavy work is now likely to be intermittent and limited'. Granted, workers still manually install infrastructure, do maintenance, and other physical tasks, but 'many of these jobs are partially or fully mechanized and much more time is spent operating machinery and driving vehicles' (McPhee 2004, p. 298). In this context, McPhee stated that machine and equipment design, and operator training and skill, are becoming more critical to successful mining. With this, there is in a sense a changed responsibility for physical ergonomics. This, in turn, comes with other problems:

> There is a ... clear responsibility on the designers, manufacturers and suppliers of mining equipment to ensure that the current lamentably low level of consideration of both the operators and maintainers of their products is significantly improved as quickly as possible.
> None of the [current problems] are subtle problems ... [and] no detailed understanding ... is needed to address what are essentially ergonomics limitations of the crudest type. The fact that such fundamental limitations can and do create serious health and safety risks shows clearly that manufactures and suppliers are currently falling lamentable short of their duty of care responsibilities.
> To redress this situation requires much more attention to be given to the consideration of human factors and ergonomics during their design processes.
>
> *(Simpson et al. 2009, pp. 26–27)*

In part, then, the consideration of physical ergonomics is outside the direct control of mining companies. However, Simpson et al. (2009) also emphasized that mining companies should take more active positions in communicating the importance of these issues. Still, there are work-design aspects that influence the physical ergonomics that go beyond the design of the equipment and machines. In terms of attractiveness, these issues relate to aspects of adequate equipment and physical activity; in attractive work there should be some type of physical workload, and it has to be healthy (Åteg et al. 2004).

Some of the most important ergonomic risks in mining come down to vibrations and manual tasks (Donoghue 2004; McPhee 2004); the consequences of these risks are a significant part of the work-related health problems in mining. With mining becoming more mechanized, the need for physical work has been reduced; in turn, manual tasks and hand–arm vibrations (hand–arm vibrations come almost exclusively from powered hand tools) are now uncommon. Today, manual tasks and hand–arm vibrations are mostly present in maintenance and activities related to infrastructure. Whole-body vibrations (WBV), on the other hand, are still a problem in mining. Almost all machinery used in mining (i.e. more or less all machines in which the operator operates the machine from the machine itself) is a source of WBV. Many neck and back injuries can be traced back to WBV. And often, WBV are caused by 'rough rides' that are in turn caused by, for example, aggressive and careless driving or poor road maintenance (McPhee 2004).

3.2.3.1 Current Controls and Solutions

Manual labour depends directly on the task in question and is more or less reduced as mechanization and automation increase. Though maintenance, which is largely manual, normally increases as automation and mechanization increase, the end result should still be a decrease in manual tasks. Therefore, one control of this hazard is further mechanization; in fact, this reasoning is commonly used to motivate further mechanization and similar initiatives (cf. Johansson 1986; Lööw et al. 2017). Other common controls are different types of lifting aids, properly designed workstations, equipment, and so on (cf. Horberry et al. 2011; McPhee 2004).

By removing manual tasks, hand–arm vibrations can also be expected to decrease. However, as the removal of manual tasks tends to result in increased mechanization, WBVs increase. Several factors influence WBV, and there are a variety of subsequent controls: the type of vehicle, its speed, maintenance, and the condition of roads are all important factors. Several ways of reducing WBV can be done through the following measures: regularly monitoring vibration levels; training operators to drive carefully; setting speed limits; making sure road problems are quickly discovered and corrected; developing effective road maintenance programs; making sure vehicles have appropriate design; making sure vehicles are effectively maintained; making sure that operators that drive a lot have variations in their tasks; and introducing regular breaks that are 'out of seat' (McPhee 2004). Additionally, measures like suspension and other machine characteristics can be used to control WBV. Like the other issues raised in this section, though to a greater extent here, vibrations are eliminated when the operators are separated from the machine they are operating. That is, where remote-control partially reduces noise and dust exposure, it removes virtually all exposure to vibrations.

3.2.3.2 Designing for Physical Ergonomics

Here, perhaps more than in the two preceding sections, the complexity and effect of the attractiveness issue becomes apparent. The first consideration is age. As noted, the current workforce in the mining industry is ageing and is not being replaced quickly enough by younger people. While older workers may not have the same physical and mental capacities as young workers, they do have comparatively better coping strategies that result in safer work practices (McPhee 2004). Age needs to be taken into consideration in the design of mining work. Additionally, as the attractiveness problem is addressed, a new population that has not traditionally worked in mining will most likely enter the workforce. It is important that workplaces and equipment take anthropometric consideration of the new workforce. For example, women are generally lighter and shorter than men. This can affect visibility (if a chair is too low for the operator, they might not be able to properly see through the windows) and vibration exposure (suspension is designed with a certain weight in mind and can amplify the vibration if the weight of the operator is not corrected for); designs need to take this into consideration.

Another issue arises in the balance of solving two different problems. As noted, remote-control removes virtually all the vibrations. However, for controlling dust and noise, isolated cabins are recommended. But isolated cabins still expose the operator to vibrations. There is no simple solution to this type of complexity, though participatory approaches appear as being a key part; McPhee (2004, p. 401) recommended 'the cooperative interchange between expert and non-expert to find satisfactory solutions to a range of problems especially where there needs to be trade-offs and compromises'.

Finally, another 'trade-off problem' is the recommendation for some kind of physical activity in promoting attractive work. Reducing strenuous physical activity tends to remove *all* physical activity. Here, however, solutions involving work rotation are more applicable as, unlike dust and noise exposure, there is such a thing as 'healthy' exposure.

3.2.4 SOME CONCLUDING REMARKS REGARDING HEALTH RISKS

In discussing health and ergonomic risks in mining, and their relation to attractiveness, we have selected a narrow focus. The purpose is not to be exhaustive, but rather to give an insight into some general problems and the approaches needed to address them; the general principle should apply to all three areas covered. None of these issues can be dealt with in isolation, and they all affect each other. This often results in trade-off situations, where the key appears to be in participatory approaches; recently, such approaches have been advocated by several researchers (e.g. Burgess-Limerick et al. 2012; Horberry and Burgess-Limerick 2015; McPhee 2004).

Another aspect is that the issues need to be approached holistically, using both technical and organizational solutions. For example, risks may first be reduced as far as reasonably possible through barriers, and then work rotation can be utilized to decrease exposure further. Additionally, McPhee (2004) argued that social contact, control, and such measures (i.e. attractive features of work) have the potential to reduce, for example, stress. He held that it is critical to match job demands with employee capabilities. Errors, fatigue, stress, and injuries may result where this is not done (i.e. when demands exceed employee capabilities). To cope with this, employees may take shortcuts that lead to unsafe practices, work at an unhealthy pace, and self-select out of jobs where they feel unable to meet the demands. Here, Carayon and Smith (e.g. Carayon and Smith 2000; Smith and Carayon-Sainfort 1989) have suggested balancing approaches. While a balancing approach is complex, it is not without reward; Carayon and Smith provided several arguments and examples of these mechanisms. But this type of balancing is not possible without the involvement of the workforce and other affected stakeholders.

3.3 SAFETY HAZARDS AND RISKS IN MINING

More than health issues, mining is readily associated with safety issues. However, while these are prominent in the industry, there has also been important improvements. In fact, one could argue that, for some mines, the description of mining as 'dark, dirty, and dangerous' is now quite far from the truth. The Swedish mining industry has gone from a lost-time injury frequency rate (LTIFR) of 51.3 in 1981 to 7.1 in 2015 (SveMin 2016). These safety levels are close to those of manufacturing and construction (Swedish Work Environment Authority 2016, 2017). Looking at fatal accidents in Sweden around the 1950s, the number per one million working hours was 1.1 (Brand 1990). Between 2000 and 2009, on average, one fatal accident occurred every other year (SveMin 2010). This represents a decrease in fatal accident rate by around 95 per cent. Similar significant improvements are also observed in, for example, the United States (Katen 1992) and Canada (Haldane 2013). In the European Union, the mining industry displayed a positive trend between 1999 and 2007 (European Commission 2010): though all sectors displayed a positive trend, mining was one of the sectors that had experienced the largest decrease in accidents during this period.

Nevertheless, while many mining countries have improved their accident rate over the years, many others continue to struggle. For example, 1384 miners died in China in 2012 (Feickert 2013) and 1444 died in Turkey in 2010 (Demiral and Ertürk 2013). In Poland, where mining is generally modern and mechanized, 311 fatal accidents occurred between 2000 and 2009 (Krzemień and Krzemień 2013). Additionally, notwithstanding significant improvements, Elgstand and Vingård (2013, p. 6) reported that 'Where reliable national statistics exist, mining is generally the sector having the highest, or among 2–3 highest, rates of occupational fatal accidents and notified occupational diseases'. Other comparisons have found that during the period between 2005 and 2008, Sweden had more than half the fatal occupational injury rate compared to Spain and New Zeeland; yet, Australia had more than half the occupational fatal injury rate of Sweden (Lilley et al. 2013). In this, it is also important to acknowledge the differences between different types of mines: Nelson (2011) showed that open-cast mines are safer than underground mines, and that underground coal mining has a notably higher accident rate than other underground operations. In other words, there remains room for significant improvements, especially as the rate of improvement has tapered off in the last decade (cf. SveMin 2016).

In a more detailed view, Table 3.2 presents the causes of accidents in the Swedish mining industry during the 1980s. It shows 'fall of a person' as the most common cause of accidents, followed closely by object-handling accidents, and other common causes being made up of strikes by falling objects, contact with machine parts,

TABLE 3.2

Causes of Accidents in Swedish Mining, 1980–1984

Cause	Count	Number of Accidents Per 1000 Employees	Proportion (%)
Electrical accidents	28	0.5	1
Fire, explosion, blasting	50	0.8	1
Contact with chemical element	70	1.1	1
Contact with heat or cold	127	2.1	3
Fall of a person	833	13.5	17
Step on uneven surface, misstep, step on nail	265	4.3	5
Other contact with stationary object	430	7.0	9
Struck by flying object, spatter, etc.	444	7.2	9
Struck by falling object	535	8.7	11
Other contact with machine part, vehicle, etc. in motion	599	9.7	12
Overexertion of body part	605	9.8	12
Object-handling accidents	706	11.5	15
Other	129	2.1	3
Total	4841	78.7	

Source: Data from National Board of Occupational Safety and Health (1983, 1984, 1985, 1986, 1987), Statistics Sweden (1982, 1983, 1984, 1985, 1986).

vehicles etc., and overexertion. The first two causes (as well as the others to a lesser extent) appear to relate to the physical and manual nature of mining work at the time. For the 2010s, as presented in Table 3.3, the most common cause of accidents by far is loss of control of, for example, machinery (the classification system changed significantly between the 1980s and 2010s, which makes it difficult to directly compare the two periods). While factors relating to manual and physical labour are still significant, machine and equipment related causes are dominant. Additionally, virtually all frequencies of occurrence have markedly decreased, except perhaps the loss of control (the first period does not readily list this type of cause). As mining has become more mechanized and automated, it is reasonable to expect that accidents involving machines will be common, especially under the assumption that mechanization and automation decrease other accidents (but see the following and Chapter 4). Table 3.3 also presents figures for the Swedish manufacturing industry. What can be gleaned from these figures is that the two industries are relatively similar in terms of proportions of causes for accidents; loss of control is a more common cause in manufacturing, while leak, outflow, overflow, collapse, fall, and breakage of material is more common in mining. This difference is likely due to environmental conditions. So, at least based on these figures, accidents in mining are at a similar level to those in manufacturing. However, with reference to the frequency rates, more or less all causes of accidents are more frequent in mining; given a 1000 employees, for every ten accidents in manufacturing there would be 13 in mining.

TABLE 3.3

Causes of Accidents in Swedish Mining and Manufacturing, 2010–2014

	Mining and Quarrying			Manufacturing		
	Count	Number of Accidents Per 1000 Employees	Proportion (%)	Count	Number of Accidents Per 1000 Employees	Proportion (%)
Electrical, explosion, fire	23	0.6	4	238	0.1	1
Leak, outflow, overflow	19	0.5	12	723	0.3	2
Collapse, fall, breakage of material	68	4.6	12	2317	0.9	8
Loss of control	206	5.0	37	14579	5.4	50
Fall of a person	106	2.6	19	4825	1.8	17
Body movement without any physical stress	29	0.7	5	1676	0.6	6
Body movement under or with physical stress	91	2.2	17	4396	1.6	15
Shock, fright, violence, aggression, threat	1	—	—	152	0.1	1
Other	8	0.2	1	300	0.1	1
Total	551	13.4		29206	10.9	

Source: Data from Swedish Work Environment Authority (2015, 2014, 2013, 2012, 2011).

Figures from the mining industry in the United States are presented in Figures 3.2 and 3.3. They show that a majority of accidents in mining are due to the handling of materials and slip or fall of persons. This is followed by machinery, but those are less than half the total accidents. This demonstrates the differences between countries, but also that modern mining needs to consider the full width of potential occupational injuries. That is, 'old' types of accident remains relevant even in new environments.

There is, of course, more to these figures than we take up in this brief analysis, but the tentative conclusion we draw here is that there is evidence to suggest that the nature of mining accidents has changed and that they may no longer be unique to the industry as such. Beyond this, while the causes of accidents and injuries in mining tell something about the safety situation in mining, they do not explain how these accidents happen. Consequently, without knowing what causes accidents in mining, it is difficult to suggest how to prevent them. Fortunately, there is research about this.

Mining operations and their environments are complex and high risk. The physical and technical environments of the mining industry is in part responsible for this situation. Hartman and Mutmansky (2002, p. 32) described the nature of mines as being due to 'falls of earth, strata gases that are emitted into the mine atmosphere,

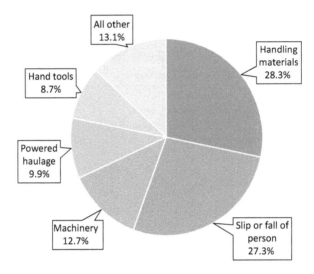

FIGURE 3.2 Distribution of non-fatal lost-time injuries by accident class in metal mining in the United States, 2008. (Data from NIOSH 2011a.)

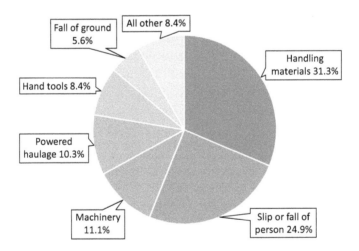

FIGURE 3.3 Distribution of non-fatal lost-time injuries by accident class in overall U.S. mining 2008. (Data from NIOSH 2011b.)

the explosive nature of mineral fuels when in the form of dust, and the many types of heavy equipment used in the mining process'. Similarly, Laurence (2011, p. 1559) observed that 'One of the reasons that hazards in mining are so great is because of the significant energies involved, be they gravitational, mechanical, chemical, electrical, or other types'. Additionally, the mining industry has mainly increased its productivity through bigger machines, bigger loads, and so on (Hartman and Mutmansky 2002). This means that the mining environment now 'contains' more energy than previously. (Also recall that Nelson (2011) identified differences between open-pit, underground, and coalmines.) The occurrence of accidents is dependent on the presence of energies (Haddon 1963) and their prevention will depend on the ability to control them (Haddon 1973).

However, this is alone is not a satisfying explanation. For example, Swedish figures (SveMin 2016) show that accidents occurring due to the mentioned energies (i.e. gravitational, mechanical, chemical, and so on) are rare: between 2011 and 2015, just under 10 per cent of all accidents were due to mucking, falling rock, or traffic. By comparison, walking, jumping, or tripping accounted for around 22 per cent, and service and repair accounted for almost 40 per cent of all accidents. Lenné et al. (2012) and Patterson and Shappell (2010) found that around nine out of ten accidents were triggered by human action (e.g. operator errors and violations). Additionally, Patterson and Shappell (2010) found that unsafe leadership was present in a third of the cases; Lenné et al. (2012) observed that, in two-thirds of the cases, organizational influences were present. Similarly, Laurence (2011) found that many accidents in mining are caused by lack of awareness or non-compliance with rules, poor communication, production taking priority over safety, inadequate training, and so on. We expand on some of these studies in the following sections but, for now, it is enough to conclude that the safety situation in mining is not explained solely with reference to the presence of harmful energies.

Through their meta-analytical investigation, Beus et al. (2016) found substantive empirical support for the idea that safer workplace behaviour is associated with fewer accidents. In turn, they found reasonable empirical support for the proximal antecedents of safe behaviour, that is, personal resources (cognitive, physical, etc.), safety knowledge, skills, and motivation affect safe behaviour. On the organizational and group level, they likewise found moderate empirical support for the distal antecedents for the proximal factors, that is, contextual factors (policies, practices, safety culture/climate, leaders, co-workers, etc.) and job characteristics (risks, hazards, job demands, etc.) affect personal resources, safety knowledge, etc. In other words, it can be assumed that changes made in, for example, a policy (a distal organizational-level antecedent) will affect personal resources, safety knowledge, and so on (individual-level proximal antecedents). This, in turn, affects safety behaviour that ultimately results in (or affects the probability of) accidents. This is to say, then, that in addition to strategies that aim to counter harmful energies, there are also strategies that correspond to distal, organization-level antecedents (e.g. policies and safety practices) that in the end seek to affect such behaviours.

In the following, we review the quite limited literature on what increases and decreases the risk of accidents in mining, as well as the nature and characteristics of those accidents. At the same time, we try to exemplify how different strategies have been and can be used to control accidents. Like the previous section, the intention is not to provide an exhaustive review. Instead, we want to illustrate what the situation in mining demands of, for example, a design process. This review is demarcated from major mine accidents and catastrophes. While, of course, a very significant topic in the operation of mines, they tend to involve many aspects that fall outside of workplace attractiveness issues, even though many factors are related. For more reading on major mine accident and catastrophes, Quinlan (2014) and Saleh and Cummings (2011) are recommended.

3.3.1   What Increases and Decreases the Risk of Accidents in Mining?

Laflamme and Blank (1996, p. 486) studied age-related accident risks in the Swedish mining industry and concluded that 'the transformation of production processes in the mine had a more rapid beneficial impact on work productivity than on accident risks' and that 'the reduction in accident rates that steadily took place did not favor all age categories of workers to the same extent'. For the first conclusion, three periods of different productivity characteristics were identified: in the first, both productivity and time worked declined; in the second, productivity increased and time worked remained stable; and in the third, there was a balance between productivity and time worked. The first two phases showed an increased level of accident risk compared to the third. For the second conclusion, Laflamme and Blank (1996) found that the changes in in work (e.g. due to technology) favoured older workers. They argued that it is doubtful that age alone explains this phenomenon and suggested that younger workers are more exposed to workload and injury risk compared to older workers (i.e. the physical and technical environment of the younger workers was more hazardous, and that they lacked experience of dealing with it) or, alternatively, that younger workers were hired when jobs did not require extensive training. This situation has a bearing on designing attractive workplaces; for example, the (attractive) changes may not affect new and younger workforce the same way it affects the older workforce.

Blank et al. (1998) studied injury risks faced by underground miners in a Swedish iron-ore mine. The study observed a 41-year period with regards to productivity and time worked. It looked specifically at measurements of risk in terms of injury per produced unit and worked hours. In terms of hours worked, they found statistically significant changes in injury rates during 'stabilization' (i.e. when productivity and time worked was stable). In terms of production volume, they found statistically significant changes in injury rates during 'striving for efficiency' (i.e. productivity increased slightly and time worked increased) and 'labor intensification' (i.e. productivity rapidly increased and time worked decreased). They argued that 'The hours-worked rate was considered as a reflection of the human efforts put into production, and therefore of individual exposure to risk; the production-volume rate was regarded as an indicators of the performance of the production system as a whole' (Blank et al. 1998, p. 270), but found that the two denominators produced similar results when the production process consumed more machine resources than human resources. In this sense, safety is improved to the extent that technology decreases risk exposure. However, they also noted that 'The extent to which this is a reflection solely of technological differences is ... debatable, since several factors might confound any relationship found between technological development and injury' (Blank et al. 1998, p. 272). Again, the prevention of accidents cannot only rely on technological developments; at the same time, future improvements in safety will need technology, but it is not enough on its own.

Blank et al. (1996) investigated technological development and occupational accidents in the same mining company but over 80 years (1911–1990). They found that three factors affected the likelihood of accidents: mechanization, reduction in working hours, and unemployment. Mechanization significantly increased the overall accident risk. Blank et al. (1996) argued that this may be because mechanization usually coincides with work intensification and work condition deterioration; or that the machines that were introduced had poor protection and were inadequately adapted at the workplace. Increased unemployment meant an increased annual accident rate that the authors suggested may be because of a decrease in labour union's bargaining power, or unemployment led to increased overtime and work intensity. The reduction of hours worked underground also increased the annual accident rate, but Blank et al. (1996) suggested that this really reflected the effects of unemployment at that period. When it comes to annual mortality rates, the study showed that automation and mechanization led to a lower rate. They concluded:

> There is a relationship between technological development and occupational accidents [but] this relationship is conditional on other factors ... taken into account, it becomes clear that changes in technology are not sufficient in themselves fully to explain variations in accident frequencies.
>
> *(Blank et al. 1996, p. 144)*

In fact, the Work Insurance Act of 1995 had the strongest negative relationship with annual accident rate; the conclusion is similar to that of Blank et al. (1998).

Outside of Sweden, but similarly, Hartman and Mutmansky (2002) argued that the improved accident rate in the mining industry in the United States between 1910 and 2000 was due to fewer fires and explosions (1910–1930); fewer employed miners and better ventilation (early 1930s); mechanization, social enlightenment, and production decline (1950–1960); and more surface mining and federal legislation (1960–1985).

Strategies to control energies and behaviour, and organizational practices and strategies, are all present throughout these studies; they have all had their role in, for example, affecting injury rates. The studies point towards such factors as time worked, hiring practices, production rate, and legislation, on the one hand, as well as new machines that offer better protection, and the move to less 'energy-rich' environments, on the other. Trying to associate these strategies with different points in time, Lööw et al. (2017) suggested that certain strategies have been more dominant at certain periods in the Swedish mining industry: early safety initiatives focused primarily on technology development; while since around the 2000s, focus has been on organizational measures, with legislation playing an important role in terms of systematizing safety practices (see Figure 3.4). The participants of the study argued that the way to reach the vision of zero accidents was to continue to focus on issues relating to safety culture, such as norms, values, attitudes, and behaviour.

3.3.2 The Nature, Characteristics, and Prevention of Accidents in Mining

Studies that have investigated accidents in the mining industry have tended to focus on machine-related accidents. Zhang et al. (2014) investigated fatal accidents in surface mining related to haul trucks. Of the 12 investigated accidents, they found some of the causes to be inadequate or improper pre-operational checks, poor maintenance,

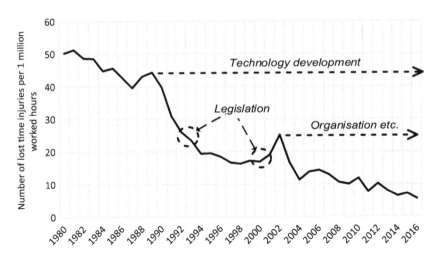

FIGURE 3.4 Accident frequency rate over time in the Swedish mining industry (data courtesy of SveMin) with initiatives for increased safety overlaid. (Based on Lööw et al. 2017.)

unstable ground, use of alcohol, failure to set the parking break, failure to follow rules, and excessive speed. In combating this, they recommended attention be given to pre-operational checks (especially since inadequate or improper pre-operational checks and poor maintenance appeared to be the most common cause) and operator training. Many mechanical failures, which in turn can contribute to accidents, are reported to be due to poor maintenance (Zhang et al. 2014). It is interesting to note here that operator maintenance, a remedy for poor maintenance, is associated with workplace attractiveness (cf. Johansson and Abrahamsson 2009).

Another study (Groves et al. 2007) looked at equipment-related injuries in general in the mining industry of the United States over a ten-year period. They found a pronounced difference between fatal and non-fatal accidents. For example, while material handling accounted for 54 per cent of non-fatal accidents, powered haulage and machinery together accounted for 51 per cent of fatal accidents. Furthermore, the type of machine involved in the accident differed. For example, the most common type of machinery involved in fatalities were haulage trucks, followed by front-end loaders, continuous miners, conveyors, (non-haulage) trucks, and cranes. For non-fatal accidents, non-powered hand tools were involved in 24 per cent of the incidents. This is three times more than the second most common machinery (rock or roof bolting machines). They offered the explanation that non-powered hand tools are larger than their household counterparts, and that they are used under difficult conditions, which increases the injury risk. They concluded that 'The prevalence of recordable injuries associated with non-powered hand-tool use indicates that significant challenges still remain in controlling these risks' (Groves et al. 2007, pp. 463–464). Again, then, it appears to be a question of controlling energies; at least to the extent that these tools are necessary for the task in question, it is an issue of protecting the worker from harmful energy. Groves et al. (2007) put forward this need for control strategies and, due to the relative frequency of powered equipment accidents, noted new development in the area of, for example, collision prevention systems. While this type of technology is indeed promising, it comes with its own set of issues (see Chapter 4).

A similar study (Kecojevic et al. 2007) aimed to thoroughly characterize equipment-related mining fatalities during an 11-year period in the mining industry of the United States. Here as well, the main recommended interventions relate to training. Kecojevic et al. (2007) referred to Kecojevic and Radomsky (2004), who found key contributing factors to this type of accident to be poor haulage road and dump design engineering as well as the failure to wear seatbelts. Essentially, then, training should only be part of the solution; improving, for example, haulage road conditions and perhaps implementing technical solutions that prevent machinery from being operated if the seatbelt is not worn, should constitute other prioritized solutions.

Ruff et al. (2011) investigated similar accidents during the period of 2000 to 2007. They drew similar conclusions to those presented above but added an investigation on worker activity undertaken at the time of the accident. To this end, they found that 25 per cent of all injuries and fatalities occurred during the maintenance and repair of machinery. They recommended that 'Innovative engineering controls should be investigated, including human presence sensors and devices that allow workers to remain at a safe distance from hazardous locations while performing maintenance or cleanup' (Ruff et al. 2011, p. 17). The first recommendation could be similar to the technologies implied by new communication and location systems, and the second is essentially different types of remote control. Again, these are promising solutions, but they are also subjected to the issues raised in Chapter 4. This is also the case for the solutions recommended for the other machine-related accidents, which primarily included different types of proximity warning systems, that is, technical solutions.

Some studies have looked specifically at contractors. Studying the mining industry of the United States, Muzaffar et al. (2013, p. 1342) concluded that 'the odds of a fatal versus nonfatal injury were nearly three times higher for contractors than that for operators during 1998–2007'. For Swedish mining, Blank et al. (1995, p. 34) arrived at similar results and concluded that

> ... a considerable part of dangerous jobs in the mining industry are performed by contractor workers. Contractors seem to get injured more often and sustain more severe injuries. They also perform other tasks and work under other conditions than mining company employees when incurring injuries.

Present in both these studies are also concerns regarding the reporting of these accidents: that accidents do not get reported or, when they are, they show up under other branches than mining (mining contractor companies are often not registered as mining companies). This means that the mining industry and contractors may be affected by more accidents than the statistics show. All in all, the 'contractor question' is important in mining for safety and workplace attractiveness. But it is complex, involving factors such as organization and flexibility (see Section 5.2.1). Muzaffar et al. (2013) recommended the following interventions to help the situation: work hour restrictions, targeted safety training, raising awareness, and reassessment of working conditions. In other words, the focus in the first three cases is on organizational-level strategies that, in the end, aim to change behaviours, whereas the fourth deals with change in, for example, the physical environment in which contractors work.

Kronlund et al.'s (1973) study is related to this type of issue but is not focused on contractors. After a mine strike at in a Swedish mining company in the winter of 1969, the wage system was changed from a piece-rate system to fixed monthly wages that varied depending on what type of work was performed. Kronlund et al. (1973) described how statistics regarding accidents developed two years after the wage system was changed. The major developments were that severe accidents decreased by 95 per cent, normal accidents decreased by 70 per cent, and minor accidents increased by 45 per cent. There were several reasons for this development, but the change from piece rates was considered to be the most important since risk-taking among the miners was reduced and sick leave due to minor injuries did no longer reduce earnings in a significant way. During work with piece-rate pay, many miners ignored injuries from minor accidents so that they would not lose any income. This shows how organizational-level strategies that are not necessarily intended to affect safety can have significant effects. (Chapter 5 investigates this type of issues further.)

Also focusing on issues of organizational and human error, Simpson et al. (2009) stated that it is a common belief that most mining accidents are due to human error, but that this is a truth with some modification. Almost all accidents (around nine out of ten) are a result of human action (e.g. an unsafe act by an operator), while the root cause is usually a combination of factors (Lenné et al. 2012; Patterson and Shappell 2010). These factors are presented in Table 3.4: organizational factors were present in up to two-thirds of mining accidents, and factors related to unsafe leadership were present in around one-third of the cases. Additionally, at times, environmental conditions affected more than half of all mining accidents: the technical environment was involved in one-third of the accidents and the physical environment in up to around 55 per cent. This is another way of illustrating the complex nature of mining accidents: it is rarely possible to point solely towards one factor. Below we detail some of these multiple factors.

Human error can be placed into one of several categories. Simpson et al. (2009) described skill-based errors as 'autopilot errors'; walking into an elevator and pressing the wrong button because another is usually pressed is an example of this. Skill-based errors are common when using PPE, tools, or equipment. For example, often vehicles are not properly parked (the parking brake might not be applied, the engine might still be running, and so on). Patterson and Shappell (2010) recommended, for example, that mining companies install alarms that activate when the vehicle exited with the engine still running or with no parking break applied. They also suggested that similar alarm solutions could be used to increase the use of PPE by fitting this equipment with radiofrequency IDs (RFIDs); certain areas would then sound an alarm if operators enter the area without the PPE. (However, this does not guarantee that the PPE is used but only that it is available, although sometimes a simple reminder can be enough. Here, we mainly want to draw attention to technical solutions that aim to affect behaviours.)

Decision errors, which Patterson and Shappell (2010) referred to as 'thinking errors' and Simpson et al. (2009) identified these as rule-based errors, are errors where the objective was to accomplish something but the plan was inadequate or inappropriate. This can include procedures that are incorrectly executed due to lack

TABLE 3.4
Underlying Reason for Accidents in Mining

Category	Study 1[a]	Study 2[b]
Outside factors	—	—
Regulatory factors	—	—
Other	—	—
Organizational influences	9.6%	—
Organizational climate	1.4%	28.5%
Organizational process	8.3%	65.4%
Resource management	1.0%	29.3%
Unsafe leadership	36.6%	—
Inadequate leadership[c]	28.3%	22.4%
Planned inappropriate operations	11.8%	33.1%
Failed to correct known problems	3.9%	—
Leadership violations[d]	1.4%	4.2%
Preconditions for unsafe acts	81.9%	—
Environmental conditions		
Technical environment[e]	35.2%	32.7%
Physical environment	39.0%	55.9%
Conditions of the operator		
Adverse mental state	12.6%	25.1%
Adverse physiological state[f]	6.3%	10.6%
Physical/mental limitations	10.8%	25.5%
Personnel factors		
Coordination and communication[g]	27.2%	15.2%
Fitness for duty[h]	0.4%	—
Unsafe acts of the operator	94.7%	—
Skill-based errors	58.9%	63.9%
Decision errors	49.0%	33.8%
Perceptual errors	4.9%	—
Violations	5.5%	57.4%

Notes:
[a] Data from Patterson and Shappel (2010).
[b] Data from Lenné et al. (2012).
[c] Lenné et al. (2012) referred to this as 'Inadequate supervision'.
[d] Lenné et al. (2012) referred to this as 'Supervisory violation'.
[e] Lenné et al. (2012) referred to this as 'Technological environment'.
[f] Lenné et al. (2012) merged 'Adverse physiological readiness' with 'Personnel readiness,' which Patterson and Shappel (2010) referred to as 'Fitness for duty'.
[g] Lenné et al. (2012) referred to this as 'Crew resource management'.
[h] See note f.

of training or instructions. Patterson and Shappell (2010) suggested that making sure the workforce has the correct training or improving the current training programs (e.g. by using more pedagogic methods) can decrease the chance of this happening. This type of suggestion as a solution is common in the reviewed research. To make personnel remember procedures after training, Patterson and Shappell (2010) recommended that tools like checklists be used. Decision errors can also occur due to situations being wrongly assessed, which could also be improved with training but would focus on identifying risks and how to act in certain scenarios (Patterson and Shappell 2010). Administrative controls, such as signs to make risks more obvious, can also be used.

Violations refers to 'bending the rules' (Patterson and Shappell 2010). Simpson et al. (2009, p. 8) highlighted the complexity of the issue:

> ... the most important aspect to appreciate in relation to violation errors is that while intentional, they are not necessarily malicious or simply a result of laziness. For example, failure to wear [PPE] may be a function of it being uncomfortable or the correct PPE not being readily available. Alternatively, failure to use the correct tool or replacement part during maintenance may be a function of availability and failure to complete all required checks ... may be a function of supervisory ... pressures to 'get the job started again' etc.

Where violations are common, one may have to re-evaluate and possibly redesign procedures and equipment that are prone to violations. It is also important that awareness surrounding violation activities be increased (Lenné et al. 2012).

Beyond human errors, around half of all underground mining accidents involves the physical or technical environment in some way (Lenné et al. 2012; Patterson and Shappell 2010). Nonetheless, even seemingly simple accidents are not without a certain complexity. Often, ground or road conditions are important factors in the physical environment. Trips, slips, and falls are the most common accidents in mining. These often occur because ground or road conditions are poor. However, the situation is not just making ground and road conditions better. Earlier, the importance of water sprays to control dust was noted, but spraying water also makes surface and road conditions worse by making surfaces slippery and muddy. As water sprays are important for controlling dust and preventing health problems, they are not easily removed with the intention of preventing another type of accident. Instead, mitigation techniques may have to be employed: boots with high traction and protection against strains can be used, and hand rails in slippery and muddy areas can be installed to prevent trips, slips, and falls (Patterson and Shappell 2010).

The technical environment concerns the design and construction of equipment. Accidents that happen because of the technical environment can, for example, be due to confusing or contradicting control layouts, such as one machine having one set of controls and a similar machine a different set of controls; if the machines are similar, one usually expects the controls to be similar as well (Horberry et al. 2011; Lenné et al. 2012; Patterson and Shappell 2010). Equipment designers are first and foremost (practically) responsible for the technical environment. However, Simpson et al. (2009) held that mining companies should take more active positions in communicating the importance of these issues. Essentially, it is important to be more active in choosing correct equipment. For example, 'standard controls' could be a demand when procuring new equipment. In cases where the design is inadequate, it is often modified on site, which, if done incorrectly, can compromise performance, safety, and so on.

3.3.3 SOME CONCLUDING REMARKS REGARDING SAFETY IN MINING

What we have intended to demonstrate here is that accidents in mining are complex and depend on both human, social, technical, and environmental factors; the solutions too must cover the breadth of this issue. The technical aspect, where the focus has long been and continues to be, constitutes one part of this. However, the impending high-tech environment of the mining industry and the safety it provides means a higher proportion of accidents will directly involve humans and their behaviours. Laurence (2011) inquired into the question of how to work with issues relating to the human factor. Citing his earlier work, he argued that more rules and regulations are not the solution, and that 'Detailed prescriptive regulations, detailed safe work procedures, and voluminous safety management plans will not "connect" with a miner' (Laurence 2011, p. 1561). His proposed solution is a framework with as few rules as possible, but that those rules are of high quality. In fact, he pointed out that the strength of the framework may not even be the rules themselves but rather the process; for example, ensuring that there is a positive safety culture as well as open and well-working communication channels. We focus on this type of process in Chapters 6 and 7.

Finally, a comment on the connection to attractiveness. The basic tenet is that work has to be safe to be attractive. But do attractive workplaces 'generate' safe work? That is, if the aim is mainly to provide attractive workplaces, will these automatically be safe? It is a relevant question because in practice it may be enough that a workplace is merely *perceived* as safe to be considered attractive. Research on 'safety space' and similar concepts (e.g. Reason 1997; see also Beus et al. (2016) and their definition of safety) indicate that unsafe organizations can be relatively spared from accidents until a catastrophe occurs. As it is likely that an employee bases his or her judgement of safety on whether accidents occur, it is possible that unsafe workplaces come to be considered safe. On the other side of the issue, it is possible to design workplaces in such a way that they are safe but also unattractive. For instance, a workplace could be designed with extensive rules and regulations with little room for autonomous decisions by employees. This could result in higher safety but low attractiveness. There is a lot to gain, then, if designing in accordance with principles of attractive work also resulted in safer working environments.

The answer, however, is both yes and no (to the extent that it can be clearly answered). On the one hand, research indicates that practices such as employment security, extensive training, self-managed teams, decentralized decision making, and high-quality work are positive elements of occupational safety (Zacharatos et al. 2005). Some of this research is specific to mining. For example, Goodman and Garber (1988) found that the familiarity between two workers in a team affected their safety, and Trist et al. (1977) found that autonomous groups had better lost-time accident rates. On the other hand, if the aim is to control energies, measures for increased attractiveness may do nothing for safety; having a good balance between work demands and autonomy does not reduce potential of harmful energy. Analogously, physical barriers to energies may do little in the way of affecting workplace attractiveness; for example, it is probably of little concern to the employee if a truck is designed to withstand one particular level of energy compared to a different level. Once again, then, these are issues that inevitably will result in some trade-off. Thus, an efficient approach to the issues should be a participatory and integrated approach: safety, ergonomics, and attractiveness need all be considered by all relevant stakeholders.

3.4 SUMMARY

In principle, this chapter concerns two things: (1) many of the health and safety issues in mining have technical solutions, some of which have been available for many years and (2) despite the existence of these solutions, accidents, and occupational diseases are comparatively common in the mining industry. As mining companies are beginning to realize, technical solutions alone are not enough to solve current problems (cf. Lööw et al. 2017); improving the stagnating trend of improvement requires the active involvement of all stakeholders. In many cases, there are no 'one-size-fits-all' solutions; an improvement in one area can introduce problems in another. The issue is to find a balance between interests and to ensure that progress in one area does not come at an unjustifiable expense in another. Our point in this chapter is that this cannot be done a priori, and that it is only through active participation of those concerned that these issues can be properly managed. Additionally, to find the best solution, the right questions must be posed. And these questions are best posed with sufficient knowledge about, for example, the causes and prevention of accidents and ill health.

4 Mechanization, Automation, and New Technology in Mining

4.1 INTRODUCTION

In mining, there are many examples of mechanization, automation, and new technology. These take many forms and produce a variety of effects. It is a topic of importance that has already received a considerable amount of attention both from a technical and human factors perspective. The latter has been covered extensively by, for example, Horberry et al. (2011), and it is not our intention to repeat this here. Rather, we continue from where these authors finished. Specifically, we attempt to draw attention to aspects of mechanization, automation, and new technology that affect aspects of an organizational nature and those that relate to workplace attractiveness. However, to start this discussion, a short review of mechanization and automation in mining is justified.

The mining industry has a long history of automation but is, in general, less developed in comparison to other industries (Bellamy and Pravica 2011; Lynas and Horberry 2011). Several aspects of mining precludes or complicates utilization of new technology and high levels of automation. Lever (2011, p. 806) noted that

> ... the highly variable and unpredictable mining environment affects the successful execution of each or sequences of unit operations. Thus, automated systems must be able to sense, reason, and adapt to this unpredictable environment in order to function effectively. ... many existing automation technologies from other industries are not readily transferred into mining.

Lever (2011) also argued that automation and robotics have yet to significantly change mining processes. Part of this is due to high capital costs and the long life of mining operations that lead to lower rates of technological change. This means, in turn, that activities in mining can resemble those used 25–50 years ago (Randolph 2011). Furthermore, the development of an underground mine can take 5 to 10 years to complete (Nelson 2011), with the implication that technology can be outdated by the time the mine is operational. Nonetheless, automation in mining is increasing and remote control is especially prevalent. Hartman and Mutmansky (2002) listed drill rigs, raise borers, hard-rock miners, loaders, continuous miners, load-haul-dump trucks (LHDs), trains, and belt conveyors as some examples of remote-controlled operations. Automated operations include trains haulage systems, hoists, drills, and LHDs (Hartman and Mutmansky 2002; Lever 2011). Although few of these are fully automated but rather classified as semi-automated.

Historically, mechanization and automation has improved the aspects of health, safety, and the general work environment in the mining industry. And improvements in these aspects serve as a strong motivator for the current automation of mining operations. Looking at mining in Sweden, fatalities and serious accidents have decreased (SveMin 2016b) during a period in which the number of people who work in mining is lower than previously but their production per employee is higher than ever before (Geological Survey of Sweden 2016). However, the effect of mechanization and automation on the work environment is far from straightforward. The effect can be small (Blank et al. 1996), difficult to determine (Blank et al. 1998), affect the workforce in different ways (Laflamme and Blank 1996), and so on. Furthermore, Eriksson (1991) observed that in addition to the positive work environmental effects, the mechanization and automation of Swedish mining during the 1970s and 1990s engendered increased lone working, division of labour, work intensification, and psychological stress. More recently, McPhee (2004) concluded that because mining work is changing, the hazard exposure is changing too. Now, he argued, risks include long working hours, fatigue, mental overload and underload, reduced task variation, increased sedentary work and work in fixed positions, and whole-body vibrations. Worryingly, he argued that there is little recognition of these risks in the mining industry; moreover, these risks will have significant negative effects if 'they are not balanced by well-informed decisions by managers' (McPhee 2004, pp. 301–302). As noted, a purpose of this book is to help managers make informed decisions – in this case, by discussing the effects of technology on social aspects.

Here it is important to remember that, owing to the fact that different unit operations lend themselves to automation in varying degrees, automation tends to progress asymmetrically. For example, Abrahamsson and Johansson (2006) studied the technological progression at a Swedish mine and classified the technology level of different operations from 1957 to 2005 on a scale ranging from a first level of manual work to a sixth of automated work (see Table 1.1). Their results show that the level of automation and mechanization is greater in the later parts of the value-creating chain; that is, the possibility for automation increases as the ore becomes more refined. As a result, 'old' work environmental problems are not solved once and for all as technological progress continues. In short, as some areas will improve, at times to levels where issues encountered are those of office environments, others will remain relatively unchanged.

The point here is that automation and mechanization may very well solve many problems but also increase the complexity of mining operations. Moreover, as this chapter aims to show, there is a complexity that is not limited to the interaction between humans and machines: it also affects mining work, its organization, and indeed the entire organization. In the next section, we focus on the effects of technology on these aspects.

4.2   ON THE SOCIAL EFFECTS OF NEW TECHNOLOGY

There is much research that has explored the social effects of new technology, but studies that focus on mining specifically are rare. Here, we review some of the important general theories that concern the effect of new technology, especially those of mechanization and automation. The kind of automation implied by digitalization and other high-tech advances is quite recent. However, many of the basic principles are general and, in a sense, timeless. An illustrating example of this is that much of the prominent research that we review was published in the 1950s, 1960s, and 1970s.

Blauner (1964) studied the relationship between technology, social structure, and personal experience as he sought to investigate the alienation and freedom of industrial workers in the United States. He proposed four levels of mechanization or industrialization (this is the classification that Blank et al. (1996) used in their study): skilled craftsmanship, machine-tending technology, assembly-line technology, and continuous process automated technology. The results of the study is apparent in the titles of the chapters: 'The Printer: A Free Worker in a Craft Industry', 'The Textile Worker: Integration without Freedom in a Traditional Community', 'The Auto Worker and the Assembly Line: Alienation Intensified', and 'The Chemical Operator: Control over Automated Technology' (Banks 1965). Peterson (1965) summarized this situation as the automobile workers being subject to the greatest alienation. With no power over the production, the workers' work activities, as well as related activities, are 'unrelated to the larger goal of production ... [and] to extra-factory institutions, and the worker's selfhood is destroyed or blunted in the process of working' (Peterson 1965, p. 83). For the textile industry, 'Conditions making for alienation are prevalent ... but they are offset by the traditionalistic integration of work with non-work concerns in the small ... towns where most textiles are manufactured' (Peterson 1965, p. 83). The printer and chemical worker have high levels of freedom (i.e. low levels of alienation). The printer is free because of the autonomy afforded by their craftsmanship. The chemical worker is free because the automated technology gave them autonomy and 'non-manual responsibility' (Peterson 1965). Blauner's (1964) conclusion is that alienation, as a function of technological level, is 'U-shaped'. For craftwork and high automation, alienation is low and freedom high; for machine tending and assembly-line work, alienation is high and freedom low. Technology, then, has the possibility to bring about both negative and positive effects. Another way to view this is that technology is not the only determining factor. Blauner's (1964) example is of objectively alienated workers not feeling alienated. This can be conceptualized as, for example, social factors, balancing out, negative factors of technology (cf. Carayon and Smith 2000). In this case, the social factors can themselves be viewed as negative, and are most likely not enough to offset the long-term negative effects of the technology. But the effect is important to note, especially as it could work in the reverse: social norms themselves may engender a negative perception of positive technological effects. This is important for attractiveness because, among other things, it connects external and internal factors of workplace attractiveness. Applying the model to mining, one finds that it used to involve craftsmanship (and to some extent still does). Machine-tending is also present; for example, miners have to supply semi-automatic drills with drill steel. It is not really possible to talk of assembly-line technology, but continuous-process technology is becoming increasingly common in the form of, for example, control room work.

Kern and Schumann (1974) aimed to chart the effect of technological development on industrial work and the relation of workers to it. Similar to Blauner (1964), Kern and Schumann (1974) classified technology and work on a scale ranging from craftsmanship to fully automated production. Pure craftsmanship was stipulated as consisting of handling of the work-piece, processing of the work-piece, control of product and work process, and correction of product and work process. In fully automated production, by contrast, all these tasks are performed by machines; the human in the system is now tasked with basic planning and control of the process. In mining, the ore can be conceptualized as the work-piece, and rudimentary forms of mining then correspond to pure craftsmanship; on the other end of the spectrum, today, some operations are fully automated, such as transport. Technology influences the work tasks. Kern and Schumann (1974) proposed that technologies give rise to six types of work tasks: craftsmanship connected to craftsmanship on one side of the scale, and control room work connected with fully automated production on the other. Abrahamsson and Johansson (2006) modified this scale for use in their study. Like Blauner (1964), Abrahamsson and Johansson (2006) acknowledged positive aspects at each end of the scale but also in some of the intermediary stages. Machine operating work, for example, can also have positive aspects, which was not really recognized by Blauner (1964).

Kern and Schumann (1974) also introduced the concept of 'range'; this concept refers to the range of work tasks integrated in a production system. It relates to the fact that the technological development creates aggregated systems where an operator may act as a sort of mediator between systems. However, as the technological sophistication increases, so does the range of the systems. This means that the operator will gradually become superfluous. In mining, the range is relatively limited. This is seen especially in the early stages of operations; for example, whereas drilling is automated, charging is not. Similarly, loading still requires an operator, while transport may be fully automated.

One of the main results of Kern and Schumann (1974), however, concerned the polarization of the workforce that occurs in the wake of automation and mechanization. In general, when technology levels increase (e.g. a manual process is mechanized, or a mechanized process is automated), some people in the workforce undertake work tasks demanding employees who are more qualified; other people in the workforce get work tasks that do not require any significant qualification. Applied to the topics of concern here, this challenges the notion that automated mining work alone will result in more attractive work. That is, it is not enough to suppose that the automation of a certain work task will engender a more attractive work environment. Even if the control room work tasks implied by automation might very well be considered attractive, there is no guarantee that other work tasks will be. The point is that all work tasks need to be made attractive, even maintenance and machine-tending jobs. This can be achieved through workplace and job design, but also through work organizational measures, such as making new work roles include both the qualified and unqualified work tasks that are normally generated by technology development.

Bright (1958) analysed and drew attention to different aspects of automation from a management perspective. He distinguished between three aspects of mechanization: the level – the degree of mechanization; the span – how many of the necessary tasks are encompassed by mechanization; and penetration – the extent to which secondary and tertiary tasks are mechanized (e.g. maintenance, lubrication, adjustment, and assembly). Bright (1958) also classified mechanization with reference to its power source (muscle or mechanical power) and the nature of control (e.g. what the operator can control and what the machine controls). An important conclusion from Bright's (1958) study is the recognition that the secondary and tertiary tasks (i.e. the penetration of automation or mechanization) are often forgotten. In effect, even industries considered highly automated, such as the process industry, contain tasks that are not automated or mechanized. This is obvious in mining. For instance, drilling is in many cases considered automated. However, it is usually only the primary task that is executed without human intervention; an operator must still transport the machine, position it, and start the process. The implication for safety, ergonomics, and workplace attractiveness is that these secondary and tertiary tasks may be unsafe, un-ergonomic, and unattractive. At one point in his study, Bright (1958) also noted the discourse of fully automated factories that manage themselves and asked readers to consider: if this is the case, why are there so many cars in the parking lot? For mining and its future, the question is whether one can rely on automation to make the mining industry less dependent on labour. Dahlström et al. (1966) considered Bright's (1958) work when discussing the effect of mechanization on qualification. From case studies, Dahlström et al. (1966) identified two possible developments: given the same type of mechanization (e.g. from low to medium, or from medium to high), work can either be further divided (lowered qualifications) or enriched (increased qualifications). This is a central issue of this book. Our standpoint is that the negative development occurs when issues of safety, ergonomics, and attractiveness are not actively considered or considered early enough. Blauner, Bright, and Kern and Schumann are summarized in Figure 4.1.

With regard to these issues, research conducted in a mining context has tended to have a more narrow focus. Abrahamsson and Johansson (2006) explored the consequences of replacing manual underground work with remote control from above ground. They found that deep knowledge about 'the rock', including the ability to 'read' it, was essential for the underground miners. For the miners above ground, it was a question of abstract knowledge, an ability to read and understand pictures and symbols and relate them to different measurement test results. Another aspect was social belonging and identity. The workers had their roots in a changing context

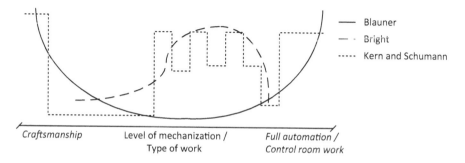

FIGURE 4.1 A rough summary of Blauner, Bright, and Kern and Schumann. The y-axis represents the investigated variable of the particular study (e.g. alienation). (Based on Helgeson and Bergman 1983.)

where they had to leave their old blue-collar colleagues and move into a white-collar environment. The old type of macho behaviour was challenged – the workers had to find new ways of forming their identity. This was visible in the workers who moved above ground preferring, at least initially, big joysticks (similar to those in the actual machines) and changing into working clothes even though they worked in an office environment. Later, however, this changed: the operators started preferring smaller joysticks and worked in ordinary clothes. The effects observed in the study need not be specific to this particular development; an important finding regarded the change in required qualification and the resistance generated by the change. The implementation of new technology has to be managed carefully and, in a sense, empathically to avoid encountering resistance that end up hindering technological development. The study pointed to challenged macho behaviours, but the effect can likely be expected in cases where, for example, unions and privileges are challenged. Even if the new technology challenges behaviours that need to be changed, as in the case of macho behaviour, conscious management is required to ensure that changes actually last.

A related point is found in Goodman and Garber (1988). While they investigated absenteeism and accidents in underground coal mining, they introduced the concept of familiarity. They argued that while there are *general* configurations of, for example, technology and the physical environment throughout a mine, there are also *unique* configurations (e.g. state of the machinery and physical conditions at the mine face) in each section. Thus, they used

> *familiarity* to refer to knowledge about the unique characteristics of particular machinery, materials, physical environment, people, and programs that exist in a particular [location] at a particular time. [The] premise is that because of the hazardous and dynamic nature of mining, [familiarity] is critical to effective production and safety practices
>
> *(Goodman and Garber 1988, p. 82)*

Goodman and Garber (1988) observed a relationship between absenteeism and familiarity. Though important, the point of note here is the relationship of the concept to several human factor issues associated with automation and new technology. Some of these, listed by Horberry et al. (2011, pp. 151–153), are poor operator acceptance of new technologies and automation after they are introduced, poor human factors design of equipment, lack of equipment standardization, a new device being essentially irrelevant to the task, inadequate operator and maintainer training and support, over-reliance on the technology by operators, and lack of physical integration of technology. These relate, at least partially, to familiarity or the lack there of. New technologies might be rejected simply because they are unfamiliar; poor human factor design and lack of standardization can contribute to unfamiliarity; irrelevant technology and lack of physical integration of technology could also be argued to be due to lack of insight into familiarity on the part of the designer, and could itself also generate unfamiliarity; and inadequate training and support could increase unfamiliarity. Additionally, familiarity is also a concept in workplace attractiveness. In any case, an understanding of familiarity is required; not only in the management of technology but in work environment management in general. The workers and operators possess this knowledge; we argue that it is through their involvement that it is possible to develop technology, and indeed workplaces, that satisfy familiarity criteria.

Beyond the social effects, it is also important to recognize the complexities and the trade-offs that have to be considered in issues of, for example, productivity (though these issues are not purely technical and also include social aspects). For example, Hartman and Mutmansky (2002) identified some dilemmas for line-of-sight remote control. Using sight and hearing, operators can efficiently operate and control a machine with less risk to their health and safety. But as they move farther away from the machine, they also lose some capacity for control, as well as potentially losing some of the health and safety benefits of operating from inside a cabin. Hartman and Mutmansky (2002) also gave examples from tele-remote operations. Here, located far from the mine face (e.g. above ground in a control room), operators still exercise manual control, but the benefits are that there is no travel time to and from the mining face. There is also the possibility of operating several machines in parallel (usually if some elements are semi-autonomous). In the latter scenario, labour productivity is markedly increased. However, it is not guaranteed that each individual machine will be more productive; on the one hand, automated machines may not perform as well as a skilled operator; on the other hand, automated machines may be able to drive faster or be subject to fewer safety regulations. In any case, Hartman and Mutmansky (2002) observed that while remote control has improved health and safety, it has not improved efficiency because the process still requires manual control. Increased efficiency, they argued, is attained only when processes are automated.

All things considered, the indication is that there is need for new technology to be managed at an organizational level. Although departing from a different theoretical background, Mårtensson (1995) proposed, with reference to work organization for human–machine systems, six requirements that largely coincides with the points brought up here. What is interesting though is that these requirements aim to ensure the performance of and in complex manufacturing systems. Below, these requirements are presented in a modified form (Mårtensson 1995):

- *Versatile job content.* Employees plan, perform, and monitor their own jobs. The jobs are a clear part of a process.
- *Responsibility and participation.* Employees or a group are responsible for the whole work task, monitor their own work, and participate in design processes.
- *Information processing.* Employees plan their own work. There is mental work, such as in new situations and problem solving. This is followed by decision making.
- *Influence on the physical work performance.* The working pace of the employees is only temporarily controlled by processes. Employees choose their working methods. Work allows for varied physical and spatial movement and motion. It is possible to leave the workplace for a short while.
- *Contact and cooperation.* There is verbal and visual contact with at least one other employee. Employees are in contact with colleagues in other steps of the process, and there is cooperation in teams.
- *Competence and development.* Employees are developed to an acceptable skill level, which is then used in tasks that are more qualified. Consequently, there is continuous training.

From these we conclude that many of the requirements for good human–machine interaction correspond to the requirements for attractive work environments. This further highlights the connection between workplace attractiveness, ergonomics, and productivity.

With the social effects of new technology reviewed, we now move to two topics of special interest. Both incorporate many of the issues raised here. The first is remote operation centres, which many of the technology developments ultimately move towards. The second is how to actually move from the mining of today to that of tomorrow.

4.3   REMOTE OPERATIONS CENTRES

Noort and McCarthy (2008) argued that future mining will rely on remote control from remote operations centres (ROCs). This will be, and to a certain extent already is, possible through the use of advanced information technology (IT); examples include automatic mobile mining equipment, automated process control, sensor technology, advanced analysis technology, and service-oriented IT architecture. Extensive cooperation will be possible through the use of computer-based systems for communication and information/data mining. This gives the mining companies better information so that well-informed decisions along the whole value-creating chain (planning, mining, maintenance, environmental surveillance, logistics and transport, coordinating of contractors, deliveries to customers, and so on) can be made. Many other future mining scenarios present similar pictures with a strong belief in automation and remote control with changed and more advanced tasks for miners (e.g. Bassan et al. 2008; Bäckblom et al. 2010); whatever the scenario, work is always conducted from ROCs.

Work in ROCs share many similarities with conventional control room work, which is a well-researched area. In the 1980s, Bainbridge (1983) challenged the classic approach of automation design by identifying several pitfalls in automation and in automating procedures, processes, and work. And her work still remains relevant today. In the classical approach, humans are regarded as unreliable and inefficient; thus their input in the control system should be minimized. But, 'the designer who tries to eliminate the operator still leaves the operator to do the tasks which the designer cannot think how to automate' (Bainbridge 1983, p. 775). She argued that this means that the operator is likely to be left with arbitrary collection of tasks, and that these tasks probably have been given very little attention. This means that humans are still needed in highly automated systems for supervision, adjustment, maintenance, expansion, and improvement. Yet, this is not always recognized, which Bainbridge (1983) argued can leave the operator with mundane tasks that have been given little thought in terms of design; that is, the operator often becomes a passive observer. However, the operator is expected to intervene when the system is not performing as supposed or if something fails. This requires manual control skills and knowledge about the process; the problem with being a passive observer is that skills deteriorate and knowledge is forgotten, so the operator might not possess the skills and knowledge required to successfully intervene in the system when needed. Bainbridge argued that it is also important that operators are skilful, so that they know they can take over from the automatic systems if required. If this is not the case, she argued, the job will turn into one of the worst kind: one that is 'very boring but very responsible, yet there is no opportunity to acquire or maintain the qualities required to handle the responsibility' (Bainbridge 1983, p. 776). In the long term, this can even affect the health of the workers. Bainbridge (1983) also identified problems in the course of automating a new process. When a manual task is automated, it is usually the former manual operators who become new operators of the automated system. These operators might perform well within the system because, having previously worked with it, they have a fundamental understanding of the technology they control. The next generation of operators might not have this understanding. Thus, training for new operators has to include process knowledge: new operators need to have the same fundamental understanding of the process as the operators who previously operated at the mine face.

Bainbridge's insights display the importance of considering this type of issue early, in the planning or design stages, and the need for involving those who are ultimately to use the system-both the current and future generation of workers. Much of this is recognized in many industries, such as nuclear power production, the chemical industry, aviation, and defence. Yet, research on work in control rooms within the mining industry is limited. Process control in mining has focused on mineral processing and not on the remote control of underground mining activities. The industry has tried to maximize the capacities of the mineral processing control systems, but this has not been managed well. One of the major reasons for this is the poor performance of control room operators who often find themselves in a difficult situation with too little useful support from the technical system (Nachreiner et al. 2006). Li et al. (2011) studied two Australian processing plants (broadly representative of typical Australian mineral processing plants) and analysed the work performed by the control room operators. They found that the control room environment was noisy (disturbing the personnel) due to the sounds from machines, people, and signalling alarms. At the same time, the workload of operators was high, especially during periods when the mineral processing was unstable. The study also found that the control operators had was mostly passive and focused on equipment. Operators tended to respond only when the process or machinery failed; there were few proactive planning activities. Equipment failure was the dominant failure of interest and little thought was spent on optimization and stabilization of the process. Furthermore, operators were often overloaded and confused by a large amount of useless data; alert signals and failure messages often came too late; and operators did not have enough time to think through what actions were most needed. Alarms were rejected or distrusted by operators because they were too complicated, but also because they did not improve or ease the operators' understanding of the different processes. The study indicated that there were significant differences in performance and knowledge between different operators, and that these differences had an impact on the stability of the production process. Most of the operators had learned by successive hands-on training with a more experienced operator; systematic training was lacking.

These findings suggest that the role of the process operators is often forgotten, even though the significance of the operators is recognized among mining professionals.

Instead, control rooms have become even more technically advanced. This requires more crucial contributions from the control room operators. The increased automation complexity has caused an increased and fluctuating psychological workload as well as the skills demanded from operators. Hollnagel (2007) thus argued that now process control system design has to enhance human capacity rather than enhancing technology capacity alone. This is not to say that system developers do not focus on human factors – there are indeed those who do (cf. Lundmark n.d.) – but there is still much to improve. The mining industry has to develop human–machine-interaction-based control systems, improve the training of operators, and investigate and improve workload as well as work organization for operators. As the previous section shows, this requires at least in part a design philosophy based on the principles of attractive work.

Miners in future mining ROCs need to be supported by intelligent systems for decision making. These systems have to integrate complex information from many functions and present the miners with information and analyses in real time, where only the most important information (deviations or problems in production) are visible at a quick glance. At the same time, operators need to have full information access all the time. Decisions must be automated to a high degree; miners will have a primarily supervisory role where they can concentrate on more advanced and complex problem solving. They will also cooperate with different groups within the mine and with external specialist teams. This can include mining engineers, logistics experts, or maintenance experts. Specialist teams will be called on whenever needed and can quickly simulate, analyse, and thereafter, adjust production. In the end, it is about creating proactive production systems. This has been on the agenda in the manufacturing industries for some time; it is time that the mining industry also begins to take this into consideration. A proactive production system is constituted by a combination of 'knowledge workers', information, and automation. Operators receive correct information at the right time, in sufficient amounts, and in the right form. A well-designed and well-functioning information system will be a necessity. Operators will have to work with planning, programming, monitoring, intervening, and learning. Again, we argue that many of these design principles correspond to requirements regarding attractive work. However, they do not satisfy every criterion. For example, work in ROCs must also ensure that criteria for spatial movement and physical activity are satisfied. In some cases, there will be situations where requirements stand in opposition to each other. While it may be tempting to talk of optimization here, there may not be an obvious way forward; it is important to consider the stakeholders and to involve them in these decisions.

4.4 THE ROAD TO FULLY AUTOMATED MINING

The majority of the visions of future mining includes automation in one form or another. Most envision highly automated and even autonomous operations. For some aspects of these visions, the technology already exists and is being used in mining. For example, remote control is quite widely utilized, even from remote locations; autonomous machines are being developed; and much of the work that modern mining entails is conducted from control rooms. Even so, the shift from the operations of today to the operations prescribed in the vision of the future will require extensive effort. What is more, while the goals might be clear, the ways of reaching them are many and subject to extensive discussion. This section explores some of the possible ways to reach the vision of fully automated mining and identifies challenges as well as opportunities within them.

In its most simple form, Hartman and Mutmansky (2002) described the process of moving from mechanization to robotics (they considered autonomy to be the application of robotics) as: mechanization leads to remote control, which leads to automation, which leads to robotics. However, the complexities of the mining industry can complicate the straightforward application of this thinking. Here, several researchers have suggested that the design process needs to be reversed, that is, that it should start with the requirements of an automated mine in relation to layout, operations, and so on, being adapted to these requirements rather than the other way around (e.g. Kizil and Hancock 2008; Noort and McCarthy 2008; see also Chapter 7). Other researchers have attempted to give a detailed picture of possible working conditions of future, more or fully automated mines. Most of these attempts display a strong belief that technology development and automation will result in a healthier and safer mining work environment in a deterministic manner; some assume progress will occur in a revolutionary manner that will completely change the mining industry almost overnight. This in spite of the fact that automation has continually developed (slowly) since the first automated truck was commissioned in one of Boliden's underground mines in Sweden in 1971. That progress will be evolutionary appears as a more likely scenario. Even so, today and even more so in the future, special attention needs to be given to the many and different aspects of automation. As noted, automation will have an impact on the working environment – the question is if it will be negative or positive. For example, Widzyk-Capehart and Duff (2007) pointed out that automation in the mining industry may lead to increased safety and productivity but that it can also lead to the opposite. The outcome is dependent on the mining industry's willingness and ability to learn from experiences in other business, such as aviation, nuclear power generation, transportation, base industries, and manufacturing.

Although the mining industry has managed to automate some of the operations, full automation is still a distant goal in most underground mines. Part of the reason for this, Noort and McCarthy (2008) argued, is because technology has not been developed with the aim of total automation in mind. Instead, focus has been on limited and local goals, such as improving safety or productivity for workers and machines (cf. the concept of 'range' in Kern and Schumann 1974; and 'span' in Bright 1958). The mining industry needs to make realistic evaluations of the commercial value of full automation. Noort and McCarthy (2008) argued that a vision of full automation must contain a total reconsideration of the basics regarding mine layout, mining methods, and sequence of development. Furthermore, full automation can hardly be justified with economic reasons alone. Instead, the driving force behind further automation will have to be the need for improved work environment and safety, and to attract mining personnel. In this respect, Noort and

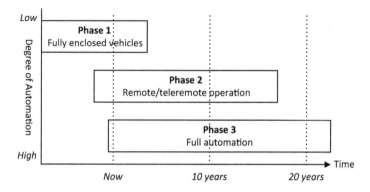

FIGURE 4.2   The model for the phases to fully automated mining. (Based on Noort and McCarthy 2008.)

McCarthy (2008) presented a model for the path to automation (Figure 4.2), which is described as follows:

- During the first phase of fully enclosed vehicles, operations like sampling, surveying, and maintenance are gradually mechanized, and operators come to control these operations from vehicles equipped with manipulators and safe cabins. The cabins resist falling rock, and the vehicles have systems that prevent the vehicle from rolling over and protect it from collisions. The safety cabin also maintains a comfortable climate. Similar safety cabins are integrated with other mobile mining-operation units, thus protecting personnel from the hostile (i.e. mining) environment.
- The second phase of remote operations means that the miners are removed from the mine front to safe control rooms where they supervise and, if necessary, steer and control different operations. No one needs to enter mining areas in production.
- In the third phase of full automation, an overarching automation system is introduced. This phase needs a total revision of the mining layouts and methods used in mining. For example, drill and blast would probably need to be substituted by mechanical fragmentation (such as a continuous miner).

Noort and McCarthy (2008) argued that the goals of the first phase are most important: the need to move further with the second and third phases is highly dependent on the degree of success in the first stage; if the first phase is successful, the need for the second and third phases will decrease. In a sense, this represents different routes to automated mining. Experiences show that the first stage is at times skipped. And there may be advantages in doing this. Instead of, for example, providing mining machines with secure cabins, mining machines are developed more in line with the second stage and forego a cabin all together. There are now also semi-automatic mining machines (such as continuous hard-rock miners) that are remote controlled, not from a control room, but from near, although not in, the machine.

Here there is a compromise in health and safety that must be discussed (see also Chapter 3). For example, while operation from a secure cabin attached to the machine provides better protection from rock falls (when they happen), dust, and noise, utilizing remote control also removes the risks associated with vibrations. Furthermore, while a cabin provides better protection from rock falls, remote control can actually lessen exposure to areas where rock falls occur; working from a cabin is more likely to take place in areas with unsupported roof, while remote control is more likely to take place in areas with supported roof. Additionally, fulfilling a similar function to that of the barriers of a cabin, by increasing the distance between the source of the hazard and the operator, a remote control option can also decrease exposure to both noise and dust. There is also the question of which option is easier to adapt to full remote control from a control room, and later full automation, in the future.

As mentioned, the mining industry can also learn much by observing the manufacturing industry, which has a long experience of automation. There, automation is a common way to improve productivity and efficiency but often tends to also create inflexible, rigid, expensive, and complex solutions. The choice of the level of automation is a delicate and difficult task as more automation is not always the right answer; manual labour remains the most flexible resource. The balance between automation/technical resources and human work/operators is therefore a central task, especially if the production is characterized by many and rapid changes. In a production context where flexibility is a principal demand, Dencker et al. (2007) suggested that proactive production is a feasible solution (cf. Section 4.3). This means that instead of simply reacting to demands and problems, the system will be far ahead and prevent problems from occurring. Such production systems integrate complex technical solutions with highly competent human operators. Dencker et al. (2007) suggested that three parameters strongly influence proactivity:

- *Level of automation.* Flexible and quickly adjustable levels of automation, applying both to mechanical/physical and information/cognitive levels of automation.
- *Level of information.* Efficient and dynamic flow of predictable and unpredictable information through the whole value chain.
- *Level of competence (among operators).* Quick, precise, and efficient competence development for the operators.

Much of these parameters touch upon issues that have already been raised. The point is, again, that there are no a priori suitable solutions, but there is a need to balance between levels.

4.5   AN EXAMPLE OF NEW TECHNOLOGY IN THE MINING INDUSTRY: INDUSTRY 4.0

An illuminating example of a new technological development happening in most industries, including mining, is *Industry 4.0*. Originally shaped by the German government in 2013 (cf. www.plattform-i40.de) as *Industrie 4.0*, it is a strategy with automation lifted to a new level through digitalization as its technical core:

technical components can be integrated, and autonomous machines can be linked to autonomous factories and, in this case, mines. *Industry 4.0* is described as the next great, fourth industrial revolution. It consists of an implementation of the 'Internet of Things, Humans, and Services' where the entire production process is included in Internet-based networks. Similar concepts exist under different names, such as General Electric's *Industrial Internet* and the Swedish government's strategy for digitalization of industry, *Smart Industry*. With introduction of *Industry 4.0* to the mining industry, one can start to talk of a *Mining 4.0*. But while this type of concept (regardless of being called *Mining 4.0*, *Industry 4.0*, *Industrial Internet*, or *Smart Industry*) can be promising, many of the points raised in this chapter are applicable to it. We therefore use this section to describe *Industry/Mining 4.0* and illustrate some potential issues.

Industry 4.0 paints a bright picture of the future industry where virtual and physical worlds will be linked into a powerful 'whole' through the integration of software from product development and production, where machines will not just do 'physical work' but perform calculations (Gill 2014; Lasi et al. 2014). This is described as cyber-physical systems, or even socio-cyber-physical systems: smart ventilation, smart logistics, smart maintenance, smart machines, and other smart systems continuously exchange information between themselves and with human workers. The German strategy highlighted the potential for skill development and a richer working life with more challenging work tasks. As seen, technological development in the mining industry stands to contribute to this. Kagerman et al. (2013) noted that it is important for companies to have a sociotechnical approach where the participation of employees in job design is central. As many of the other examples have shown, without participation, the desired effects may not be achieved at all. Several commentators have argued that *Industry 4.0* requires a flat organization with more organizational innovations, learning, enhanced human–machine interaction, and a more human-centred view on the new technology (cf. Dombrowski and Wagner 2014; Kopacek 2015; Lee, et al. 2014). At the same time, the German strategy also emphasized that the need for employees will be reduced as functions such as remote controls and preventive maintenance will inevitably reduce labour costs as well as increase the security of the remaining factory staff (Kagerman et al. 2013; Lee et al. 2014). Making employees redundant can be problematic. However, as the mining industry is struggling to recruit people in the first place, this might be a desirable development (in that it lessens the dependence on a future workforce). Here, some (especially unions) argue that competence development is crucial. For the unions, it is an insurance that the workforce can change work tasks rather than having the workforce be replaced. For mining companies, it means the current workforce can be employed in new tasks. This means a new workforce does not necessarily need to be recruited.

Analogous to its application in an industrial context, *Mining 4.0* conceptualizes the miner as an expert who ensures that production runs smoothly and that is no longer 'locked' in a control room. Instead, real-time process data and the status of machines follow the miner as they move around the mine. The miner can solve problems on the spot by remotely interacting with other operators, experts, suppliers, and customers in multi-competent teams. Production control could even be done in a digital model far away from the mine. In short, *Mining 4.0* envisions an augmented miner who has extended 'senses' and memory through technology. This technology takes advantage of and supports human skills and increases situational awareness through, for example, sensors embedded in the clothes of operator, while keeping an uninterrupted operational vigilance.

Romero et al. (2016), using the technical core of *Industry 4.0*, formed a typology of the future operator, *Operator 4.0*. This is built on eight characteristics that can be seen as the core of the new technology; we have modified them to relate to the future miner:

- The super-strength miner (physical interaction) using biomechanical support for increased limb movement, increased strength, and endurance.
- The augmented miner (cognitive interaction) using augmented reality (AR) for integrating information from the digital to the physical world. Examples here include maintainers receiving direct assistance from equipment manufacturers. Through special glasses that send and receive live video, both parties would be able to see the problem. It could then be solved through instructions from the equipment manufacturer.
- The virtual miner (cognitive interaction) using virtual reality (VR) for simulation and training of real situations that might contain risks. In fact, VR training is already relatively common in mining (Horberry et al. 2011) and is probably one of the industries where it sees the most application. Applications include training for high-risk scenarios, such as a fire event, and simulating new equipment (how will the new machine look, will it fit, etc.).
- The healthy miner (physical and cognitive interaction) using wearable sensors for monitoring health-related metrics as well as GPS location. Again, these are developments that to a certain extent have already taken place in the mining industry: already there are advanced positioning systems in use, and there have been projects on the application of sensors for monitoring miner health and safety.
- The smarter miner (cognitive interaction) using intelligent personal assistants for interfacing with machines, computers, databases, and other information systems. Currently, there are examples where RFID tag systems are used together with a smartphone app to rapidly and easy report malfunctioning equipment.
- The collaborative miner (physical interaction) using collaborative robots for performing repetitive and non-ergonomic tasks. In a sense, this is represented by semi-autonomous machines in mining, such as semi-autonomous loaders: the operator first 'teaches' the loader the path to take, while controlling the machine remotely during the actual loading. Here, the repetitive and ergonomically strenuous task of driving to and from the muck pile is taken over by the machine.
- The social miner (cognitive interaction) using enterprise social networking services for interaction between operators and between operators and the Internet of Things. The introduction of underground Wi-Fi and 5G has made this possible. And while most mines do not have special enterprise

social networks, the access to Internet in mines also means an access to social media.

- The analytical miner (cognitive interaction) using big data analytics for discovering useful information and predicting relevant events.

This classification points to the numerous possibilities of integrating *Industry 4.0* with human labour – some good and some bad. Note, though, that this development is not about new jobs being created. Rather, it is a development that means that most current jobs will be influenced by these characteristics and developments.

The visions of *Mining 4.0* not only change the technological landscape of mining workplaces and organizations, but they also cause a qualitative knowledge transformation from bodily and tacit into more abstract and theoretical knowledge and skills. In the optimistic view, *Mining 4.0* can be envisioned as requiring workplace learning as well as continuous education and systems that make use of the workers' skills, that is, a learning organization. Using Kern and Schumann's concepts (1974), there could be a clear transformation from the craftsman-like qualification into more technical qualifications. The new demands for team work, responsibility, and comprehensive understanding of production flow can be seen as a movement from qualifications dependent on the process to qualifications more independent of the process (cf. Kern and Schumann 1974, 1987; Bright 1958; Blauner 1964; Johansson 1986). What was earlier the tacit knowledge of the worker will be formalized into theoretical knowledge, digitalized, and used in computers and smartphones. In this transition, there are contradictory movements of upskilling (rapidly changing skill demands and more theoretical, comprehensive, and communication tasks) and deskilling (fragmentation of individual craft knowledge and whole tasks) (Abrahamsson and Johansson 2006). This also implies that there may be polarization: some people in the workforce are upskilled, and other people are left the same or deskilled.

In the optimistic visions of *Mining 4.0*, smart systems, automation, and remote control will take over dangerous as well as routine work so that operators can focus on learning, creating, and valuing work tasks in a safe environment (Gill 2014). Even if the development will not engender the same outcomes as the positive visions predict, depending on how the new technology is developed and interpreted, there will most likely be new types of mining work and new types of work environments; thus, there will also be new work environmental problems. For example, digital technology and remote control, together with the emerging global and sometimes boundaryless work, not only results in increased freedom to decide how and where to work, but also results in higher demands of availability. This change may blur the boundaries between work and private life. Moreover, since the ability to control and monitor the individual increases, there will be a risk of new psychosocial stress. An increased information flow and accessibility could also lead to anxiety and job strain (Hoonakker and Korunka 2014).

Mining 4.0 not only means almost invisible automated systems but also workplaces with robots and interactive machines inhabiting the traditional workspace inhabited by people. Digitalization also represents an opportunity to empower humans by providing information that builds knowledge and understanding and gives them 'super powers' and possibilities for wider communication (Feki et al. 2013). One step in this direction is information systems built on sensors that are worn by humans, the so-called wireless body area networks or body area sensor networks. These systems could be useful as a preventive safety system for operators working in harsh environments (Alam and Hamida 2014), where mining is an obvious case. At the same time, one must be aware that these systems are a threat against personal integrity.

There are many questions that must be cleared along with the introduction of *Mining 4.0*. The development cannot and should not be stopped, but it requires reflection and consideration so that more problems are not created than are solved. It is important that the mining industry is active in this work, but we also know that it will take time and there will be much traditional work environment work required before this becomes reality.

4.6 CONCLUDING REMARKS: NEW TECHNOLOGY DOES NOT AUTOMATICALLY SOLVE PROBLEMS

In concluding this chapter, we want to bring to attention some concrete problems associated with implementing new technology in mining that intends to solve issues relating to, for example, safety and workplace attractiveness. And we especially focus on both drawbacks and advantages of each new solution. This summary is based on Lööw et al. (2017).

The first example regards the implementation of battery-powered LHDs. One of the reasons behind developing and using these machines is to increase safety. As noted in Chapter 3, the current safety situation in mining is partly due to the energies involved. In this sense, the new technology does not represent a marked improvement because there would not be a significant change in the energies involved. For the more secondary effects on safety, an illustrating example is the reduction in noise that battery-powered vehicles entail. High noise levels have several negative effects on safety as it can mask warning signals and negatively affect situational awareness (e.g. it may not be possible to hear oncoming traffic). High noise levels can also produce stress reactions that affect attention and comfort, which can also increase the risk for accidents. On the other hand, however, battery-powered machines that are more silent can make other personnel less aware of them. This could increase the risk of accidents. Another aspect is that maintenance requirements is expected to decrease with battery-powered vehicles. Repair and service activities accounted for close to 40 per cent of all accidents in the Swedish mining industry during 2011 and 2015 (SveMin 2016b; cf. Ruff et al. 2011). The decrease in or removal of maintenance could therefore increase safety. Finally, results from a focus group suggested that battery-powered machines are *perceived* as less safe (Jäderblom 2017). Regardless of actual safety, perceptions of safety is itself important for safety performance (cf. Nahrgang et al. 2010). Any implementation of the technology must ensure, on the one hand, that safety levels remain the same or improves, and on the other hand, that operators and potential employees are equally convinced that it will.

The other technology that was investigated in the study was semi-automated chargers. These are also motived by increased safety. The idea is that robotized charging decreases risk exposure by moving the operator away from the most hazardous areas. While safety will increase in this sense, the effect on overall safety may be different. For example, lone work was expected to increase and has increased in the past following similar technology implementation (see Eriksson 1991). Lone work can increase risk (primarily by increasing the consequence of an accident). Furthermore, Horberry et al. (2011) identified, for example, over-reliance on technology, inadequate training, behavioural adaptation and risk homeostasis, and general poor equipment design as potential issues when implementing new technology. These issues all influence safety. Another consideration is that rock fall is responsible for comparatively few accidents. Only just under 4 per cent of all accidents in Swedish mining are due to falling stone (specifically, between 2011 and 2015 only two serious accidents in Swedish mines were due to rock fall; SveMin 2016a). Meanwhile, repair and service activities accounted for close to 40 per cent of all accidents (SveMin 2016a), which could be problematic considering automation may increase the need for maintenance and service.

New technology will not solve all the problems. Much of the new technology is good – or has the potential to be – but it requires conscious management, analyses, and active participation before being fully ready for implementation. This is also means that while, in the end, several work environmental issues will be solved, in the interim, and probably for a significant period in the future, 'traditional' work environment management and activities are required.

5 Work Organization and Management in Mining

5.1 INTRODUCTION

Good workplaces, employee well-being, safety, high performance, and productivity depend on good work organization and management. In fact, most issues raised in this book relate to organizational issues at some level. Even so, it appears that the mining industry does not have the same tradition or experience of dealing with work organizational issues compared to other industries (Cavender 2000; Fiscor 2014), and more frequently turns towards technology to solve problems. Additionally, Hutchings et al. (2011) noted that management practices in the resource sector is under-researched.

Work is always organized in some way, even if much thought has not been put into the issue. The problem is that if no or little effort is spent on work organization, there will be several negative outcomes. Furthermore, while the work organization the most relevant only once the organization is active, many decisions in development, planning, and design stages affect work organizational issues. That is, work organizational practices such as group work can only be practiced once there is some human activity. But decisions regarding, for example, technology will influence the extent to which group work can be practiced. Our aim with this chapter is to review some of the more prominent work organizational and management issues in mining. However, this area is difficult to delineate, and many issues have in part been raised in earlier parts of the book. In an attempt to cover issues that the other chapters miss, this first section takes a wider approach to capture these issues in a more general perspective. The second part of this chapter proposes a management approach to several of the topics of the book based on a common management concept.

5.2 WORK ORGANIZATIONAL ISSUES IN MINING

This review builds and expands on previous research that we have published (cf. Johansson et al. 2010a; Johansson et al. 2018). In the following sections, we have updated and revised some of the issues to be more in line with this book and current trends.

5.2.1 CONTRACTORS AND FLY-IN/FLY-OUT PERSONNEL

Contractors are common in mining and over-represented in accident statistics. Generally, contractors perform different tasks under different conditions than regular employees. Moreover, accidents involving contractors are more severe and occur more often (Blank et al. 1995; Muzaffar et al. 2013). However, the issue of whether to use contractors or not is more difficult and complex than it may first appear. In the discussion of sustainable mining societies, mining contractors may be favoured instead of in-house personnel: when mining operations cease, a number of contractors remain in the society. Optimally, these contractors have broadened their activities to other industrial sectors so that they are able to continue with their activity after the mine has closed down. The alternative, using in-house personnel, would mean the workforce is not equipped for work in other sectors. Another aspect that speaks in favour of the use of contractors is the development and ownership of competence. Contractors with a certain variety of activities have the opportunity to gain experience in different contexts that they can transfer and use productively in mining. A problem is that if knowledge is not possessed by the mining company (in this example), it may be more difficult to integrate in an efficient production development process. Other positive aspects of contractors includes external companies bringing new ideas and working methods into mining operations that can lead to the implementation of new methods of work organization. Not only can this increase overall safety but also makes use of the contractors' potential for innovation and efficiency. In fact, as these companies can provide opportunities for variety in terms of work assignments or specialized tasks requiring high levels of competence and skills, working for a contractor may itself be considered an attractive option (Johansson et al. 2018).

Extensive use of contractors, however, especially where contractors themselves utilize sub-contractors, introduces a risk of an 'us and them' culture emerging as workgroups from different companies are active in proximity to each other (Nygren et al. 2017); this can negatively affect both safety and workplace attractiveness. Indeed, an important factor that speaks in favour of in-house competence (i.e. to not use contractors) is the possibility of systematic health and safety activities. As indicated by experiences in Sweden, Poland, and Finland, it is difficult for mining companies to maintain equal and good safety practices when many different contractors are engaged (Johansson and Johansson 2008). Additionally, some mining companies that have relied extensively on contractors have found that they have lost control over competence; they no longer possess core competencies and, in some cases, are entirely dependent on contractors to provide it for them. This can impair technological development and introduce risks in, for example, mine production.

A related issue is whether to live on-site or use fly-in/fly-out (FIFO) solutions; for the individual, it is an issue of choosing where to live, and for the company, which option to work towards. Labour supply is determined not only by the work itself and the working environment; the society and what it can offer its residents is of equal importance. One reason to stay in the place where one lives is satisfaction with the local context, where social relationships with family, friends, and relatives, opportunities regarding work, education, housing, leisure activities, and so on impact such decisions. These aspects are important in issues related to the attractiveness of the mining industry but, in some cases, can also affect workplace level factors. For example, access to leisure activities and closeness to family and friends may influence stress tolerance.

A starting point for the attractive workplace is that people work in a prosperous community and to as far as possible avoid a so-called FIFO society. In the discussion about the social responsibility of mining companies, these companies are expected to construct a strong technical and social infrastructure that ensures the survival of a society after the mining has ceased (Ail and Baffi 2007; Wibowo and Rosyid 2008). Extensive FIFO practices decrease opportunities to meet this expectation. However, it is unlikely that an individual mine alone can support the infrastructure of a society at a level required to ensure that a community is attractive; ensuring sustainable community development requires a diversified economy. A well-functioning community structure is difficult to establish in small, remote mining towns. For maintenance, one could presume a number of tasks can be solved by the existing staff as long as they have access to online support from suppliers. But there are scenarios in which FIFO organization might be required. In such cases, it becomes important to consciously manage this workforce.

5.2.2 Wage Systems

A well-motivated workforce is a prerequisite for high productivity; the use of different types of piece-rate wages to increase and maintain good work motivation is widespread practice in the mining industry. However, the effects of such systems are disputed. Johansson et al. (2010b) found that in many situations, piece rates have a negative effect on health and safety. This review of piece rates identified several studies that showed the negative effects that they have on different aspects of health and safety. Although research is sparse and fragmented, much of the accumulated knowledge about the effects of piece-rate work reviewed by Johansson et al. (2010b) shows that in many situations piece rates have a negative effect on health and safety.

There is also evidence to support this hypothesis from the mining industry. After the Swedish mine strike in 1969, the wage system at LKAB was changed from a piece-rate system to fixed monthly wages, which varied depending on what type of work that was performed. Kronlund et al. (1973) described the effect of this change on accidents in the two years after it had been made: severe accidents decreased by 95 per cent, normal accidents decreased by 70 per cent, and minor accidents increased by 45 per cent. Several reasons for this development were identified, but the change from piece rates to monthly salaries was considered to be important because risk-taking among the miners was reduced and absence caused by sickness due to minor injuries no longer reduced income in a significant way; during work on piece-rate pay, many miners ignored injuries from minor accidents so that they would not lose income.

5.2.3 Working Hours

The practice of extended workdays (regular shift lengths of 10 or 12 h/day, while still maintaining a 40h working week) is common in the mining industry. It appears to be a popular solution among workers due to the increased number of days off compared to traditional schedules. There is fear, however, expressed by management, workers, unions, and experts on occupational health and safety, that working long shifts will increase the risk of occupational accidents and health problems. Some research confirms these fears, and some does not. For example, Cliff and Horberry (2008) studied hours of work as a risk factor in coal mining. Their results did not indicate any strong association between hours worked and the number of incidents or injuries, and different forms of shift work did not display differences in regard to severe injuries. Other studies, however, show clear negative effects of long working hours. Dembe et al. (2005) investigated the impact of overtime and long working hours. They used a large survey and analysed the responses of more than 10,000 blue collar workers in the United States. The results showed that working at least 12 h/day was associated with a 37 per cent increased hazard rate, and working at least 60 h/week was associated with a 23 per cent increased hazard rate. Dembe et al. (2005, p. 595) concluded that 'These results are consistent with the hypothesis that long working hours indirectly precipitate workplace accidents by inducing fatigue or stress in affected workers'.

For future and modern mining, then, it is important to be aware of the described and obvious risks and, if possible, avoid overtime and extended working hours. Employers and trade unions have a shared responsibility to consider the risks when closing agreements about shift forms and working hours.

5.2.4 GENDER

Gender and gender equality is an issue of major importance to the mining industry's economy and its ability to attract and recruit new qualified employees. For example, during a period of 10 years, a particular Swedish mining company made investments in gender equality and diversity interventions. This was accompanied by a clear and rapid increase in the number of women working in the mine: from 5 per cent in 2005 to almost 15 per cent in 2015. The company in question has promoted gender equality since the 1960s, with examples including the retraining of women and recruitment campaigns targeting women; the analysis of the work environment to identify positions that it is possible for women to hold; an annual pay review; and removing erotic pictures from locker rooms. Furthermore, the company established women's networks, gave gender-awareness training, ran recruitment campaigns for executives and technology experts, and introduced an educational programme focusing on mining at upper secondary school level (with 50 per cent of attendees being young women). Today, it is important for the company to depict women as integral to modern mining as part of its marketing strategy. For example, the company presents itself as a modern high-tech company; on its official website it is stated that the company has a trademark based on (among other things) gender equality and diversity, where gender equality is viewed as an important aspect of the company's work with social sustainability and social license to mine. In combination with a society that is responsive towards these issues, the gender equality and diversity interventions of the company appear to be displaying positive results. At least the number of women that work at the mine is increasing, as is the number of women who are production managers. Additionally, policies for the organizational culture have been changed.

Nonetheless, there are accounts that differ from this positive image. Eveline and Booth (2002) gave an example of the disappearance of women from mining work in an Australian mine; the proportion of women decreased from 28 per cent in 1984 to merely 4 per cent in 2000. Eveline and Booth noted that women had to cope with subtle sexism as well as open hostility and sexual harassment; there was open opposition to practical arrangements, such as toilets for women in work areas, the provision of smaller-sized safety gloves, and ideological opposition to the recruitment and advancement of women. Some practical jokes generated physical dangers for women. While the situation in Swedish workplaces is generally better, women report comments that suggest that women do not belong in the mine, that the work is too dangerous, too unhealthy, too demanding, or too technical for women.

A related aspect is a macho-masculinity culture that makes the work of mining companies in relation to safety more difficult. In some mines, there exists rather nonchalant attitudes towards safety, such as employees working over open shafts without a safety device and not reacting to the alarm on gas indicators. Another example from coalmines in Australia is a workplace culture, observed by Abrahamsson and Somerville (2007), which was so strongly founded on risk-taking, competitiveness, and 'macho-masculinity' that it acted to suppress many years of safety training and education that the company had provided to its employees. Andersson (2012) noted that men at a mine in Sweden often took risks because of solidarity with other men, a form of moral obligation between them. In an underground coalmine in Canada, Wicks (2002) observed similar connections between masculine institutionalized identities and organizational dysfunction, which created a situation in which a fatal accident occurred. There are also reports that risk-taking is perceived as semi-sanctioned by some of the management: talk may be of 'safety first' but when it really comes down to it, reality is different.

Another aspect of this macho-masculinity culture is resistance to change. There is a tendency for miners to exaggerate the nature and characteristics of the old technology, the difficult work environment, the dangers, and the risks of the work with the intention of greatly emphasizing the mine as a dangerous place. Mining labour unions also draw attention to the hazards of mining, both at a general level in the mass media and internally in negotiations with the employer as an argument for more safety, better working conditions, and higher wages. This is hardly surprising given that a large part of the high income miners receive is legitimized on the grounds of the special conditions of underground work. If safety is perceived as too good, and mining as too much like ordinary industrial work, it will probably be difficult to legitimately argue for high wage levels. The image of dangerous mine work also functions to protect the status of the work and retain the old identity of the 'real miner'.

Abrahamsson and Johansson (2006) gave an illuminating example of gender-related resistance. In 2005, the first front-loader operators (all men) were moved from an underground location to steer and control the loading machines remotely from the remote operations centre on the seventh floor of an office building. Initially, employees continued to use work-wear and changed clothes in the locker room after every shift, despite the fact that they were just as clean as when they arrived. This stopped after a year or so when they began to wear jeans and t-shirts or similar clothing, just as others who work in an office do. Abrahamsson and Johansson (2006) explained this as a desire to socialize with the employees in the locker room environment. They gave an alternative interpretation of this being an attempt to be 'real' miners, to be the same as those who work in the dirty and more dangerous underground environment; it is difficult to maintain the image of macho mine work in a control room. Abrahamsson and Johansson (2006) also argued that the resistance can be explained as an effect of remote control, that the work tasks seem much simpler after they have been moved out of their context (e.g. the physical place where the loading or drilling machines are situated). When the work tasks do not need to be conducted in a mystic, difficult, and tough work environment, the aura of 'job secrets' and tacit knowledge or craftsman-like qualifications disappears. This means that it can be difficult to retain the feeling that mine work is unique, and the connection to the history of the mine, thus generating resistance.

From a macho-masculinity perspective, it is not only new technology and safety that are problematic. With an increasingly wider variety of people working in the mine, it becomes more difficult to maintain a feeling of uniqueness as mine workers and links to stories of past mine work. In relation to this, gender equality initiatives are also being met with resistance and negative attitudes, such as 'There are enough women in the mine now' and 'You cannot sit beside a woman on the bus without being accused of sexual harassment'. The new technology is perceived as an enabler for more women to be employed at the mine in the future, especially when work tasks are moved above ground and controlled remotely.

Despite these problems, changes are already happening. Other things are important today, as compared to yesterday. People who work in mines are simply adapting to them. The old type of macho-behaviour and the male mine workers' social system is progressively becoming obsolete. Employees need to move on from the old blue-collar worker roles and take a position in a high-tech, or even white-collar worker, environment. There is an increasing number of women employed as miners and as managers. Moreover, the workplace climate is changing to include more positive and inclusive attitudes.

From this we sketch two scenarios. The first contains changing mineworker masculinity to suit this new situation; mining work continues to be just as manly as before, merely different and connected to a different form of masculinity. The numerical male dominance can be sustained. The increased number of women in mining does not create gender-mixed workplaces; gender segregation continues but in a newly created form (cf. the polarization effect of Kern and Schumann; see Chapter 4). The second scenario is that mining work is 'purged' of masculinity: the link between mine work and masculinity is removed by de-gendering to create mine work that can be performed by both men and women: an occupation in which gender does not matter. The aim must be the latter scenario.

One of the hopes of technological development in the mining industry is that it will allow for changed gender patterns; a better work environment combined with higher qualification demands will enable more women to work in the industry, creating better gender equality. But the picture is not so clear-cut. Here, we use digitalization as an example (see also Section 4.5). At many industrial workplaces where digitalization is taking place, it is quite common that technology is associated with masculinity (Berner 2003; Mellström 2004). This masculinization of technology is evident in the discourse of the technology as well as in the company culture. At traditional male-dominated industrial workplaces, such as mining, even if the workplaces undergo digitalization, the connection to masculinity lingers because of the old strong symbolic links to a traditional blue-collar masculinity (Abrahamsson 2006;

Andersson 2012; Lahiri-Dutt 2007, 2012). As noted, mines are often associated with explicit expressions of a special type of masculinity, 'macho-masculinity'. Here men, more than in other workplaces, find it difficult to be associated with competences, attitudes, or behaviours that are seen as female (Gherardi and Nicolini 2000; Ely and Meyerson 2008; Somerville and Abrahamsson 2003) or have associations with unmanliness (Connell 1995). As a result, there is a seemingly paradoxical tendency that workplaces and work tasks introduced as a result of automation, computerization, and robotization can undergo a process of 'feminization' while the men hang on to the old technology (Olofsson 2010). One example of this is when the underground mine workers, half-jokingly, gave the remote-control workers sitting above ground nicknames such as 'the velour workers' (Abrahamsson and Johansson 2006; Andersson 2012). On the one hand, this trend opens up for new view on gender in industry; on the other hand, this trend can be seen as a symptom of a conservative organization, that is, barriers to implementing the new technology, and are therefore important to study and understand.

5.2.5 SOME ADDITIONAL NOTES ON WORK ORGANIZATIONAL ISSUES AND WORKPLACE ATTRACTIVENESS

While work organizational issues raised above are important in their own right, they also have important bearing on workplace attractiveness. Sometimes this connection is quite explicit, but other times it is more subtle. In this section, we try to bring out some of these aspects.

As noted, working as or for a contractor may in itself be attractive: it might be that the contractor company manages to offer benefits that the local mining company cannot, or that it entails working in a professional and close-knit group. The problem in terms of attractiveness is not the contractor companies themselves; as long as these companies manage to provide attractive work, the arguments of this book are not different for these companies compared with other mining companies. However, the use of contractors might be problematic in the hiring companies. For example, requirements for social contact and relations can be complicated by extensive contractor use: it might be harder for the hiring company's workforce to get to know contractors. Depending on the situation and context, then, the use of contractors can be positive or negative for workplace attractiveness.

Closely related to this discussion is the FIFO issue; many of the points made for contractors apply here too. Moreover, looking at attractiveness criteria of localization, there are additionally trade-offs to consider. Mining often takes place far away from even smaller population centres; especially where the expected life of a mine is short, it is hard to justify establishing a community. FIFO might be preferred where it facilitates better access to community, services, and so on and thus contributes to making mining work more attractive.

The practice of FIFO is often combined with the practice of extended workdays. This and the question regarding wages bring to light a problem of non-advisable practices being preferred be employees. In this case, extended workdays are sometimes preferred over other arrangements, and there are also instances of performance-based wages being favoured. Additionally, for work to attractive it often has to facilitate flexible worktime arrangements. In Sweden, there are even an example of a trade union being in favour of performance-based wages, whereas almost every other union is not. Here is a situation where, on the one hand, one practice increases attractiveness, and on the other – as we can expect safety to be negatively affected due to the practice – decreases attractiveness. Evidently, this is a difficult issue that enters into a territory where simple participation of the workforce is not sufficient. In cases such as these, it might be necessary to further increase the scope of consideration in decisions regarding the design of mining workplaces to include the surrounding society. In this, there is also the aspect of whose preference is actually being considered and taken into account to consider: when we, for example, say these practices are preferred, we mainly refer to the current workforce. The problem is that the mining industry must expand its view on its potential workforce. That is, while the current workforce might prefer these practices, the new workforce may not. It is because the objective is to attract new workers that we talk of widening the scope of consideration. The discussion on gender entered partly into this as well; if asked, the masculine men of the mining workplaces would most likely, as was seen, prefer things to stay the same, but this is the problem we are trying to address!

5.3 RATIONALIZATION STRATEGIES AND MANAGEMENT CONCEPTS IN THE MINING INDUSTRY

There are no simple solutions to work organization issues, and no one intervention or strategy will solve all the problems. However, we argue that there are approaches that at least address many of them. Here, we have conceptualized such an approach as 'lean mining', taking a cue from 'lean production'. To this end, this section first gives an overview of lean production and some of the critique levelled against it. Next, we discuss how a conceptualization in mining might look, and how it addresses problems of a work organizational nature.

5.3.1 A Short History of Lean Production

Lean production originated in Japan in the mid-1980s and builds on a Japanese production philosophy. During this time, Japanese production technology and work organization practices were seen as a saviour for Europe's industry. One of the first terms to be embraced was *just-in-time* – one of the fundamental concepts of Japanese production philosophy (Schonberger 1982; Zappala 1988; Ohno 1988). It is the idea that all material should be in active production and not be allowed to accumulate inventory costs. Products should be manufactured on demand instead of being 'pushed' through production via a well-designed plan. The object was to produce and deliver 'just in time', including components and pre-assembled components. Manufacturing should only be steered by orders from internal and external customers, that is, customer demand based production. According to this method, each station orders the manufacturing of goods from the previous station, which in Japanese plants was often conducted using a simple order card, a *kanban*. The result is a more flow-based layout where the machines are either placed in conveyor belt-based flow groups (work stations) or in the form of conveyor belts.

Another tool in the philosophy is 5S (*Seiri, Seiton, Seiso, Seiketsu*, and *Shitsuke*, normally translated as 'sort', 'straighten', 'shine', 'standardize', and 'sustain'). The object is to engage every employee in all aspects of production and, with orderliness, create an efficient and conducive workplace. The aim is also to gain an overview and make production, the flow, and any shortcomings visible so that improvements can be made. Supervisors or team leaders should be instantly able to ensure that everything is in place and that functions correctly.

Continuous improvement, or *kaizen* (Imai 1986), is another Japanese philosophy in which organizations should continuously make every effort to improve on every last detail, that is, develop existing, stable, and standardized processes in small steps. The object is to develop processes and flows that allows one to 'get quality right the first time'. Each work station should have its own built-in quality control system to quickly remedy any fault or problem when it arises.

Total productive maintenance (TPM) (Nakajima 1988) is a vital accessory both in Japanese production philosophy and lean production with clear relations to TQM (total quality management; see below). Its purpose is to create disruption-free production by encouraging all employees to get involved and continuously make small improvements and preventative maintenance (preventive maintenance is in itself a common practice of lean production). The basic idea is to integrate operational work with maintenance work by having operating personnel perform some of the daily servicing and simpler maintenance work on the machines that they operate.

Standardization of tasks is a key part of Japanese production philosophy. It is regarded as the foundation of *kaizen* and built-in quality assurance work. Occasionally, job descriptions (i.e. standards) are made in such detail that a 1-min task can be broken down into 28 different stages. Part of this is known as *pokajoke*, which literally means 'idiot-proofing'. This entails designing technology and tools to be so simple that the operator literally cannot make mistakes in positioning, number of operations, operating sequence, and so on. This is used partly as a means to train employees, but it also enables management to check whether the job descriptions are followed: if the task, production technology, or even the whole flow needs to be modified. Standardization is also applied to administrative and development work; product development and technology can also be standardized.

5.3.1.1   The First Wave of Lean Production

Lean production (Womack et al. 1991) is an American concept that is largely rooted in the Japanese tradition, particularly the Toyota system. While originally an organizational model for flexible, conveyor belt-based mass production intended for the auto industry, the ideas of the Toyota system have spread to other industries with different production set-ups and are today thought of as lean thinking (Womack and Jones 2003).

Lean production is a system of fully rationalized production that focuses not only on the workforce and capital but also on capital that is tied up in products. Using the methods above, production flow is developed and managed so that every component is in the right place at the right time and with the right quality. Flow-through time is a central concept in lean production, where sometimes an entire organizational change focuses on flow-through time.

As lean production systems have mainly been implemented by industries with mass production and conveyor belt (or similar) manufacturing, its tasks are repetitive, short-cycle, micro-managed, and standardized. But standardization is also a key concept in its own right: there are strict disciplinary requirements when it comes to time, costs, quality, and safety across all parts of the organization, in all departments and at all levels; the organization is hierarchical and bureaucratic, has several levels, and there are numerous team leaders and departmental supervisors. But implementing lean production does not mean that fundamental organizational structures need to be changed to any great extent: it has mainly been used as a comprehensive set of organizational tools, principles, and work methods for management and control, though they are normally regarded as being part of the 'totality' and should be combined with a change in company culture.

In the Toyota system, there are seven types of waste or *muda* (Shingo 1981). Waste does not contribute the product's refinement or value and should therefore be eliminated; removing slack and reducing waste is at the very core of lean production. It is every employee's task to continuously observe and help eliminate waste, which are of seven types: (1) overproduction, (2) waiting times, (3) unnecessary transportation, (4) unnecessary processing or reworking, (5) inventories (e.g. intermediate storage), (6) useless motions, and (7) scrap, repairs, and inspections.

In some cases, lean production becomes a modified version of the ideas of the Taylorist system where time discipline, standardization, and micro-management are key. 'Work smarter not harder' is one of the defining concepts of lean thinking, and in practice this could be a manifestation of Taylor's theory of 'best method'. 'Not harder' is, moreover, stretching the truth somewhat given that work productivity is one of the cornerstones of lean production. The central idea is that organizations should review and streamline all parts of the production process, including the shortest breaks or work motions. The proponents of lean production argue that the main difference between it and Taylorism is that with the former, personnel are viewed as a wholly necessary part of the teamwork, continuous improvements, and quality assurance work. This is why lean production is sometimes described as *democratic Taylorism* (see Adler 1995). Womack et al. (1991) underlined that organizations with no slack are dependent on the employees not only being disciplined but also knowledgeable, flexible, and having the ability to get involved and take responsibility within certain boundaries. Proponents of lean production do emphasize that removing all slack or 'waste' from the organization can be problematic: if taken too far, there is a risk of overloading both personnel and machines, which can lead to quality and safety problems. A similar problem is that there is a risk that strong customer demand-based production creates irregularities in production volumes or in the range of products; this should also be avoided as they cause waste. It is more important, therefore, to have a balanced, plannable flow rather than maximize production in short bursts.

5.3.1.2   The Second Wave of Lean Production

Lean production has been shown to be one of the most resilient and pervasive organizational concepts. But the version being discussed and implemented today is different from the one that existed in the 1990s. Even though it still embraces conveyor belts or similar production methods, as well as the methods covered previously, lean production is now seen more in terms of a cultural transformation and philosophy (cf. Womack and Jones 2003; Liker 2004; Hines et al. 2004). It has an unmistakable holistic approach that encompasses the totality of the organization, where value-enhancing chains are highlighted (Womack and Jones 2003). The focus is on development processes, vision, and long-term thinking. Additionally, an eighth waste category has been added: the 'unutilized creativity of the personnel' (Liker 2004; Womack and Jones 2003). It mentions the importance of widening the learning radius to form a shared learning network of partners and suppliers.

Staff motivation, commitment, and learning are often cited as tools to create a base for innovation, flexibility, and a way of taking advantage of the individual's full resources as part of the production system. Lean production proponents talk about the importance of extending workplace learning into a collaborative learning with a network of partners and suppliers. The organizational form that is proposed is based on multi-functional teams, workplace learning, empowerment, and diversity. The dedication and motivation of the personnel is key. The object is to facilitate innovation and flexibility. In general terms, it is a concept that has developed into a synthesis: it seems as lean production has acted as a sponge and absorbed many aspects from the other management models, current societal trends, and theoretical perspectives. It is possible to see, for example, the inspiration of the socio-technical principles and human work science.

5.3.1.3 Strengths and Weaknesses of Lean Production

The international management literature paints a bright and optimistic picture that assumes that when companies actually implement lean production (or other current management concepts), it will result in better productivity, survival in the global market, and at the same time, good working environments by getting rid of classic Taylorism. Bad work environment is waste according to lean philosophy. The standardized work operations are seen as a necessity both for planning the production flow and for continuous improvements, including the work environment (Berglund 2006). By providing workers with learning opportunities and empowerment – for example, in *kaizen* activities – it is hoped that lean production can help workers improve their competence and also create working conditions that combat stress and work-related health issues. The expected effects of the time savings in lean production are always presented as something good: better quality and more time for innovation, development, and even improvements to the environment thanks to lower energy and material consumption (today, the term 'green lean' is also being used; see Kurdve 2014). Moreover, gender equality is on the agenda for both unions and employers, mostly when recruitment and human resource management is discussed. Gender equality is also seen as a natural part of lean production, or at least a prerequisite: as a tool for the successful implementation of lean (Franzen et al. 2010). This line of argument builds on the idea that the process and flow organization of lean production, at least in its theoretical and ideal form, requires a focus on skills and flexibility, and therefore a gender marking of work tasks becomes obsolete and contra-productive (Abrahamsson 2009).

In spite of the high level of consensus, the diversity of optimistic perspectives, and the extensive dissemination of the ideas of lean production throughout industry, many companies run into 'restorative responses', diffuse resistance, and even open conflicts when trying to implement lean production or other similar organizational models (Abrahamsson 2002). But even when change processes run more smoothly, how much of the ideals of lean are reflected in industrial implementations appears to be an open question. There are successful examples that produce good working environments (Hunter 2008; Saurin and Ferreira 2009), effective organizations, and positive financial numbers (Vink et al. 2006). But unfortunately, several negative effects have also been observed. Some researchers have even proposed the label 'mean production' as a substitute for lean production (Berggren et al. 1991). Some researchers describe the effects of lean production in terms of anorexic work organizations (Radnor and Boaden 2004) or seductive and greedy organizations where workers are expected to be disciplined, flexible, responsible, multi-talented, and prepared to work overtime (Rasmussen 1999). Others argue that much of Taylorism remains in modern organizational forms, especially in lean production, and can continue to result in the same types of work-related problems (Thompson and Warhurst 1998).

There are several contradictory depictions when it comes to teamwork, worker autonomy and decision-making, job rotation and expanding work assignments, learning, and working environment. There are examples where the perspective has shifted from the group to the individual again. The group or team is no longer the smallest planning level for employers. The cycle time of assembly lines is often not as short as during its heyday, but they are considerably shorter than with group-organized work. In general, the leaner the production becomes then the greater the risks for one-sided work movements and physical over exertion (Landsbergis et al. 1999; Westgaard and Winkel 2011). In addition, repetitive standardized work assignments make breadth and variation difficult to achieve and do not provide good opportunities for learning (Thompson and Warhurst 1998; Ellström 2006). Such situations can result in workers feeling that they are replaceable and not qualified (Wärvik and Thång 2003). Moreover, standardized work tasks considerably reduce worker autonomy and opportunities to make decisions (Jürgens 1997).

In a similar way, there is an obvious discrepancy between rhetoric and practice when it comes to some aspects of gender equality. Even if implementation of lean production *can* challenge or even threaten, as a secondary effect, many aspects of local prevailing gender inequalities, it often does not. Taylor (2006) argued that lean production does not change the unequal gendered work organization and employment forms at all. Other studies have shown that lean production intensified prevailing gendered divisions of labour and disproportionately affected women (Newsome 2003), limited effects of favourable job stress to men and limited possibilities for job rotation (Angelis et al. 2011), and created more gender-homogeneous jobs and pronounced gender segregation within the production organization (Losonci et al. 2011).

From this, we conclude that depending on how the lean production concept is implemented, it can offer opportunities for both productivity and improvements in the working environment; however, there are pitfalls. Therefore, there are different interpretations and attitudes to these systems and their claimed outcomes and effects on the work system.

5.3.2 TOWARDS LEAN PRODUCTION IN MINING

While lean production is present in virtually every industry (cf. Bhamu and Singh Sangwan 2014), it has yet to be fully implemented or embraced by the mining industry. Cavender (2000) and Fiscor (2014) argued that the mining industry has limited experience with management concepts and industrial engineering techniques in general. We hold that the mining industry cannot be said to practice lean production in the same way as other industries, and part of the reason may be that there are incompatibilities between the industry and the concept. To some extent though, lean production is being practiced by mining companies. For example, Claasen (2016) found that 35 per cent of operating mines in South Africa claimed to be practicing some lean production or just-in-time principles. Other researchers (e.g. Khaba and Bhar 2017) have recommended that the mining industry adopt the concept.

As noted, there are both positive and negative aspects of lean production. Among the positive are several effects that could contribute positively to workplace attractiveness and even safety. But there are also many effects that would work opposite to this. In adopting lean production to the mining industry – that is, to conceptualize lean mining – there are two tasks: (1) to match the appropriate principles and tools of lean production to the mining context and (2) to do so in a way that avoids its negative effects.

5.3.2.1 Conceptualizing Lean Mining

One approach to conceptualizing lean mining, presented in Lööw (2015) and Lööw and Johansson (2015a, b), is summarized here. Value is central to lean production but could be expanded beyond its usual definition in lean mining. Wijaya et al. (2009) argued that mining companies have several indirect customers, including stakeholders such as society, government, and media. The values and opinions of these stakeholders could influence the mining company's definition of value. An example of this would be taking into consideration society's 'green' values by ensuring that ore is produced with as little environmental impact as possible. Because a mine's potential impact on society and the environment is much greater than that of an automotive factory, this broadened view on value could be particularly important.

Similar to lean production, standardized work is desirable for lean mining (Dunstan et al. 2006; Hattingh and Keys 2010; Wijaya et al. 2009; Yingling et al. 2000). However, standardized work is more difficult to apply to mining than to traditional manufacturing (Haugen 2013; Wijaya et al. 2009; Yingling et al. 2000) due to the variations inherent in the mining environment (standards usually require predictability and stability). Thus, standards intended for mining activities have to be more flexible than 'traditional' standards (Hattingh and Keys 2010; Yingling et al. 2000). Wijaya et al. (2009) discussed the possibility of standardizing rock bolting by basing the number of bolts used and the pattern utilized on 'worst-rock' conditions; while this would increase the material used and time taken, it would reduce variation in process time and improve quality.

The notion of TPM can be useful for mining and is important in lean mining. Because of the long distances involved in mining and expensive machinery, downtime can be very costly. Some mines suffer from extensive machine downtime

(Haugen 2013). The utilization of TPM could decrease machine downtime. It means that operators are taught additional skills that create broader work roles. Training is essential in TPM because if the operators that are supposed to perform the maintenance do not know how, more uncertainty and variations will result from faulty repairs and inadequate maintenance (Wijaya et al. 2009).

As quality is hard to influence in mining activities, the focus of lean production on quality is not as applicable to the mining industry. However, quality in supporting or auxiliary functions is possible (Haugen 2013; Wijaya et al. 2009; Yingling et al. 2000). Haugen (2013) talked of 'internal quality', which is the quality of infrastructure. Here, low quality results in uneven ramps or pillars. In turn, low quality would result in instability. Safety is also considered to be part of the quality of supporting functions (Wijaya et al. 2009; Yingling et al. 2000). For example, focusing on quality in supporting functions could ensure that rock bolts are installed correctly, shotcreting is done correctly, or faces are only being worked under safe conditions (Yingling et al. 2000).

As with lean production, work in lean mining is recommended to be organized into teams (Yingling et al. 2000). However, this can be complicated because the machines used in mining tend to be designed for one person. Furthermore, an operator is usually assigned to only one machine (at least for each shift). Although a case study was able to show the potential of organizing work in teams (Klippel et al. 2008), this was done in a mine with a lower level of mechanization. Yet, group work is practiced in some tunnelling and development projects; these activities can be organized in small teams (Haugen 2013). Historical accounts also serve to illustrate the adoption potential. Teamwork in mines with a high level of mechanization may be a question of what level of control is exercised. Team-based organization in lean mining could mean control that is exercised at a group level rather than at an individual level (cf. Johansson and Abrahamsson 2009). For example, a group may be assigned a face to work and tasked with delivering a certain amount of ore, but which workers are assigned to which machines and for how long is left entirely up to the group.

The topic of multi-skilled workers is important to lean mining (cf. Haugen 2013; Helman 2012; Klippel et al. 2008; Yingling et al. 2000) and lean production. Mining still involves a lot of work with machines designed for one person. Therefore, even if work is not organized in teams, multi-skilling is still important because lean mining requires a flexible workforce capable of operating different machines (as opposed to only one or two as is often the case today) (Yingling et al. 2000). To some extent, this already appears to be the case. Operators in some mines are already multi-skilled (including having knowledge about maintenance), sometimes to the extent that it is not possible to trace which operator performed a certain task at a given time (Haugen 2013). Because of this skill set, operators can rotate to get variation in their work and reduce stress. There are also times when an operator does not know what task they are to perform during their shift before the shift starts (Haugen 2013). This combines well with a team-based approach to work. The competence and training of operators is also important (Dunstan et al. 2006; Haugen 2013; Helman 2012; Steinberg and De Tomi 2010; Yingling et al. 2000); at the very least, this is something that is required for operators to become multi-skilled. Operators should be trained and educated in the basics of time studies and ergonomics as well as being able to utilize basic analysis principles (Yingling et al. 2000). It is even suggested that operator training could replace work standardization (Haugen 2013). This argument suggests that if operators are sufficiently trained, standards become redundant as the operator would know the best course of action in most situations.

Supplier integration in lean mining could take the same form as it happens in lean production (the integration of suppliers happens at a high level of the organization so the type of industry concerned is not as important). Yet, this concept could have one addition. Though the integration of suppliers is not normally related to contractors, they should be included in this practice. Contractors are important in lean mining due to the fact that some mining companies are already dependent on them (Elgstrand and Vingard, eds. 2013) and that contractors' working environments and accident rates are worse than those of regular employees (Blank et al. 1995; Muzaffar et al. 2013). Lean mining has been used to solve problems associated with contractors (Castillo et al. 2015) or increase their performance (Dunstan et al. 2006). However, Wijaya et al. (2009) noted that engagement from contractors in practices such as 5S and continuous improvements might be more difficult as the contractors do not necessarily share the same values and investment as the ordinary employees.

The biggest difference between lean production and lean mining is the inability to directly apply flow-based principles to central mining activities (Haugen 2013; Mottola et al. 2011). These practices require significant modification before they can be utilized in the mining industry (Maier et al. 2014). Some of the reasons for the incompatibility include the 'tradition' of constantly pushing production (Haugen 2013; Yingling et al. 2000), the practice of manning expensive machines at all times to justify investment (Haugen 2013), long distances and the tendency for large batch operations (Wijaya et al. 2009), and variation in production process (Haugen 2013). However, supporting functions should be capable of adapting these practices.

5.3.2.2   Is Lean Mining the Same as Attractive Mining?

Attractive mining should be safe. Lean mining focuses on safety but through quality rather than 'non-entry'. That is, safety in lean mining is mainly a question of quality (cf. Haugen 2013; Wijaya et al. 2009; Yingling et al. 2000). This is because of one of the unique characteristics of mining: workplaces are 'constructed' in order to produce ore (and even as ore is produced). This means, for example, that a focus on quality could ensure that all drifts and tunnels are sufficiently rock bolted and shot-creted. In this sense, lean mining is less dependent on technology for safety. Also related to safety, continuous improvements and their related activities could also be used to systematically manage the work environment (see also Chapter 6).

Additionally, in attractive mining workplaces, management is supportive of and appreciated by their personnel, and there is cooperation between these two parties. Lean production does not only allow for cooperation but it is required between management and personnel. Successful implementation of lean production requires managers to act like coaches rather than bosses. However, there is always the risk that efforts such as continuous improvement and engaging the workforce remains only in spirit but never acted upon, and that the influence of the worker is never really enacted. Furthermore, there is the risk that these kinds of practices only become a way of accessing worker knowledge. This can lead to management no longer seeing their true value as problem solvers. Therefore, mining under a lean mining philosophy has to be built on agreement, consent, and cooperation between management and personnel. An example of this would be that management clears obstacles for the workforce, which continuously improves its work, workplace, and the business; the operators are the true experts on the process.

Attractive mining and lean mining both advocate group-based production. However, the current level of mechanization complicates this practice. Future technological development might facilitate or hinder this practice. As machines get more advanced, it may only be economically (and practically) viable to train a few operators. This would hinder group-based work. Technology may also develop in such a way that cooperation between different 'functions' is required for optimal operation (e.g. work in a control room). This would facilitate group work. Alternatively, a solution might be to exercise control at a group level rather than at an individual level. Even though the work may remain individual, it is organized at a group level. In either case, group-based production requires a skilled and authorized workforce that can go beyond the relatively narrow framework that most work organizations offers. The mining industry does not lack experience in teamwork, but to succeed in developing and implementing a lean production-based philosophy, this must be further developed; mining production is a complex activity that requires a skilled workforce that is able to resolve the large variety of problems that may occur, and lean mining could contribute by advocating a workforce of autonomous, goal-oriented work units with a high level of competence.

Attractive work requires that there is learning and flexibility in mining. Learning, especially generic theoretical learning, will create flexibility. This is consistent with lean mining. Under a lean mining philosophy, every team would be able to operate every machine and be capable of repairing and maintaining the machines that they operate. However, it is not enough to just give the teams these responsibilities. Operators have to be given formal training, both for operating and repairing the machines. Resources are also required that allow the operator to take the necessary responsibility. Failing to do so can result in productivity losses, downtime, and inactive operators. On the other hand, succeeding in this can result in production being more flexible and productive. The lean mining operator's role should be broad, challenging, and include a holistic perspective of the production process (all of which are required of attractive mining workplaces). The vision is for future mine workers to be employed simply as operators rather than trucker drivers, repair and maintenance personnel, and so on.

But a lean mining philosophy alone is not enough for attractive mining. While it offers promise in providing productive and competitive operations that at least partially considers the 'socio-technical' aspects of work, lean mining also risks resulting in developments that are opposite to what is desired. For example, issues relating to gender are not adequately dealt with in lean production practices. Furthermore, lean mining does not automatically ensure that working hours are flexible and based on social needs. Stress and psychosocial issues are another pressing matter. Lean production has been shown to increase stress and contribute to work-related illnesses. Rationalization, in this sense, would come at the cost of increased stress. Still, while these represent weaknesses of the concept, it does not actively stop positive developments in these areas.

With this in mind, lean mining – or other management concepts that relate to high performance work systems – can still be considered important for the design of future mines and mining work. For example, designing with a lean mining philosophy in mind would mean that when new equipment is procured, the choice should be the machine that best allows for work in teams (all other factors being equal); when a mining method is chosen, the method that allows for the most flexibility and is most in line with the company's values (which by extension would also be those of society) should be selected; and when a new technology is introduced, whole production teams should be instructed in its operation so that flexibility and learning is assured.

This discussion is specific for lean production and mining. But the points raised here are applicable to several other management concepts, both current and those that come in the future; management 'fashions' and practices come in waves and tend to include practices and principles of other contemporary concepts or of those that came before it (Røvik 2000). Our discussion of lean mining should thus be seen as an example of how other management concept could come to affect mining as well. Again, we want to bring attention to the need for *conscious* management. Mining companies must look beyond initial and major effects to also consider less obvious 'side effects'. Without such considerations, the management concept can end up exacerbating some of the issues it set out to solve.

6 Management of the Mining Work Environment

6.1 INTRODUCTION

So far in this book, we have focused on the 'what', that is, what are the problems of safety in modern mining. We have also tried to indicate the principles for working with and solving these problems as well as how to make use of the opportunities offered in the context of modern mining. We have yet, however, to present methods and approaches that concretely deal with the issues concerned. The purpose of these two final chapters is to present such methods and approaches. We do so in two stages. First, in this chapter, we are chiefly concerned with how the work environment can be managed in operational stages, while in the subsequent chapter, the focus is on development: how ergonomic, safe, and attractive workplaces can be planned and designed.

In this chapter, then, we focus on work environment management and some tools that can be used in this task. We first describe the principles behind, and the development of, work environment management in general, together with some applications in mining. Then, we cover methods for analysing safety and the work environment. Finally, some indicators that are useful to use in this analysis are described.

As a general principle for this chapter, we are not aiming for detailed descriptions of the available tools and methods; there is already plenty of dedicated literature available for it. Neither are we aiming to give a definitive answer regarding which tool or method to choose. Instead, we keep our discussion at the principle level, where we use some specific tools and methods to exemplify these principles.

6.2 WORK ENVIRONMENT MANAGEMENT

Work environment management, in its essence, more or less entails managing different risks or occupational health and safety issues (i.e. risk management and occupational health and safety (OHS) management respectively). Usually, a distinction is made between risk management and work environment management (i.e. occupational health and safety management): risk management deals specifically with the management of risks; work environment management deals with all potential work environmental issues (though the term work environmental *risk* is usually used). Here though, we do not consider this distinction in any greater detail. We argue that work environment management must include risk management and that risk management depends on how the work environment is managed. We make use of the wider term because the issues of concern affect a wider rather than a narrower spectrum. Of course, risk management can be defined to include a wider set of issues such as risks of accidents, production problems, ergonomic problems, additional costs and material damages, organizational problems, and inefficient work environment work (Harms-Ringdahl 2004). However, the term work environment management is still preferable as it is less readily associated with more specific areas.

By including risk management in our preferred term, we also include risk analysis. At a glance, it shares many principles with work environment management and its distinction from risk management is not always entirely clear. However, we view risk analysis as one of the tools of work environment management. We also treat it separately from work environment management (even though it constitutes a part of it) because it is also a tool that can be used outside of this process. Next, we cover the wider 'management framework' and follow it up with some available tools (or rather, the idea behind them).

6.2.1 What Is Work Environment Management?

Frick and Wren (2000) stated that, traditionally, authorities have provided companies and managers with 'recipes' on how to comply with regulation. Harms-Ringdahl (2004) noted problems with this approach. For example, he argued that for at least 100 years, there has been awareness about the squeezing-injury risks associated with industrial machinery, and there is plenty of supporting material (provisions, handbooks, etc.), but still this type of accident continues to occur frequently. On the one hand, he concluded that there is a strong belief in standards such as CE markings. These, however, are lacking in terms of risk analysis (i.e. the machines that are CE marked are not sufficiently risk analysed). On the other hand, he argued that sometimes the provisions (i.e. the so-called recipes) are hard to follow; for example, judging if they are applicable or not, or how they are to be implemented, is complicated. There is also the problem of the sheer number of provisions; according to Harms-Ringdahl (2004), there were 12,000 available provisions and similar requirements in Sweden at the time of that publication. This affects the ability of a company to deal with all of them. Thus, work environment management strategies represent an attempt to turn the process 'upside down,' making companies and managers assume a proactive responsibility (Frick and Wren 2000). Frick et al. (2000, p. 6) argued that this type of strategy recognizes

> ... that it is very difficult – if not impossible – to specify every detail of how OHS conditions should be improved at the workplaces and then impose these requirements upon employers. ... there has also been a growing push for increased 'self-regulation' of OHS, where the details of the problems and appropriate solutions are defined at the workplaces.

(We do not in any way intend to suggest that provisions and laws should not be followed – the employer must always regard and follow relevant demands and laws! Working through a work environment management approach, however, should contribute positively to an employer's ability to do so.)

Work environment management as a strategy has steadily gained in popularity since the 1980s and has evolved into one of the main concepts for reducing social and economic problems of ill-health at work, even extending to developing countries (Frick et al. 2000). It follows that there are several different approaches and conceptualizations of work environment management (again, our intention is not to argue for one particular solution but to review important aspects of the concept; we hold that the varying mining contexts are too diverse to argue for one specific approach). Indeed, some conceptualizations recognize the significant variability of industrial settings in general and therefore seek to provide general principles rather than detailed criteria. Here, Frick et al. (2000) made a major distinction between two general divisions: *systematic* work environment management and work environment management *systems*. It is essentially a distinction between regulated and voluntary work environment management: mandated principles for systematic work environment refers to the former and comprehensive work environment systems refer to the latter. More precisely, Frick et al. (2000, p. 4) phrased it as the distinction between 'the (principally simple) regulation of *systematic* management ... [and] the voluntary and (usually) highly specified (i.e. formalised and documented) ... *systems*'. They also highlighted a distinction in employee participation between the two perspectives: the mandated forms tend to include the right of employees to participate (to be informed, be consulted, and make decisions) in the management of the work environment; most voluntary forms focus more on top management, where employees may be consulted but are seldom involved in actual decision making. Our standpoint throughout this book is extensively in favour of employee participation, including participation in decision making (see Chapter 7); this is comparable to how others (e.g. Horberry et al. 2011, 2018) have argued in favour of user- or human-centric design of mining equipment and technology.

Beyond the differences, there are also similarities. Frick and Wren (2000, p. 18) listed the following core principles of systematic work environment management: (1) work environment conditions are an aspect of production; (2) therefore, top management is responsible for these conditions; (3) top management has to integrate work environment considerations into all other management decisions; (4) systematic assessment and evaluation of the work environment is essential to its improvement; and (5) work environment interventions require tasks and resources to be adequately distributed. The first principle, which we have frequently discussed, is the notion that production and work environment depend on each other (i.e. a good work environment creates conditions for good production, and the other way around). In mining, more than elsewhere, safety can be argued to be more integrated into production because tasks such as rock reinforcement forms a part of the production cycle (cf. Lööw et al. 2017). In this case, it is also clear that sufficient rock reinforcement also creates conditions for good productivity. The second principle is in part dictated by law or specific standards. The third principle we have also highlighted at several points: the notion that decisions relating to the work environment and the development and upholding of safe, ergonomic, and attractive workplaces is influenced by the entire organization. This can be choice of mining method, decisions on shift work, the physical location of the mine, and so on. The work environment must factor in all these cases. The fourth principle is essentially what we are concerned with here, that is, systematic work environment activities. The fifth is somewhat outside the scope of this book but relates to the so-called infrastructure of work environment: Frick et al. (2000) suggested that the work environment infrastructure should be able to support and develop work environment activities through information, knowledge, and facilitate perceiving positive outcomes.

In addition to this, Frick et al. (2000) held that a work environment management *strategy* needs to be established with reference to the company's surrounding context. This requires a climate that accepts different interests and supports mutual learning between stakeholders: 'A structural context supporting legitimate, productive, interest-based dialogue must be developed' (Frick et al. 2000, p. 14). This is especially important in the discussion in this book as it does not deal with work environmental problems in the traditional sense. Rather, they are problems that require changes in line with, for example, the preferences of those currently outside of the workplace, to create workplaces that a *new* workforce is also interested in working in. Chapter 7 is more focused on this issue; for now, note that Frick et al. (2000) argued, to this end, that a strategy for systematic work environment management can provide a process in which all involved parties continuously improve their ability to take up new areas of relevance (e.g. safety, health, and well-being), devise preventive solutions, support decision-making integration, and adequately anticipate and cope with a changing context.

Work environment management, then, especially in the extent that it relates to risk management, is concerned with decision making. Ideally, decision making weighs in all relevant factors and then finds the optimal decision in consideration of cost, effects, advantages, disadvantages, and other aspects. This was suggested by Harms-Ringdahl (2004), who also noted, however, that this is indeed the *ideal*: decision making is normally a much less rational process that seeks to satisfy short-term and immediate needs, owing to a lack of time, knowledge, and resources, or that a too specific or demarcated goal is pursued. A similar line of reasoning provides the motives to discuss a participative management process. Harms-Ringdahl (2004) held that risk management is seldom the simple and objective identification, assessment, and treatment of risks. Instead, it is a negotiation and interaction between different actors and their interests and values: the individuals/employees, the organization/employer, and the society/institutions. Within an organization, there are different goals, demands, and so on. For example, there might be conflicting goals: should the focus be on production, safety, the external environment, or customer demand? External pressures might introduce demands that are in contradiction to internal visions and goals; unions and management may have different views on issues and make different demands; and so on. Virtually no decision is made that only affects, or that de facto regards, only one goal; the large variety of goals makes decision making significantly complex (Harms-Ringdahl 2004). Here, the issue of lack of information regarding, for example, what the goals of different actors actually are, resurfaces; the management process in which this decision making takes place must act to take into consideration these goals and identify them in the first place.

6.2.2 WHY SYSTEMATIC MANAGEMENT?

As noted, the systematic management of the work environment is not the only way to work with these issues. And while it is hard to evaluate the effect of systematic environment management, there are several arguments in its favour of its practice. For example, Frick and Wren (2000, p. 41) argued that 'Experience in the management of OHS and product quality control, demonstrates that it is easier to prevent the creation of OHS quality problems than to detect and abate them afterwards'. To proactively identify and deal with this type of issue, there must be an underlying systematic approach. Frick and Wren (2000) also noted that adopting work environment management systems may yield advantages such as being a way to attract competent personnel. They made comparisons with the ISO 9000 standard and how market demands have forced employers to become certified according to this standard. This can lead to companies being better able to sell their product and may have positive effects on their brand, such as being viewed as a safe employer (by consumers who are more concerned with this type of issue). Additionally, the implementation of work environment management in itself may increase work attractiveness in certain cases: Hedlund and Pontén (2006) concluded in their study that the proper implementation of systematic work environment management can help improve workforce recruitment.

However, note that the 'systemization' of work environment management in itself does not necessarily produce good results (hence the call for 'proper' implementation). For one, the relevant conditions may be outside of the employer's control (Harms-Ringdahl 2004). Additionally, it is in a sense possible to systematize bad practices. It is therefore important that the goals of the management process are suitable. These goals – what work environment management is to achieve – are in turn dependent on the organization's maturity. Here, Frick and Wren (2000) distinguished between three levels that correspond to the maturity levels of quality management (again showcasing the connection between quality and the work environment; see also Section 5.3.2). At the first level, focus is on so-called numerical specifications, such as the number of lost-time injuries. The goal at this point is to reduce the given numbers below specified levels. At the second level, Frick and Wren (2000) argued that it is about 'getting things right'. They gave as an example that machines have to have safeguards and that exposure levels cannot be exceeded. At the third and final level, the authors argued that the workers' wishes and needs for safe and sound work must be satisfied, which requires that goals be set locally and consciously in dialogue with them. These maturity levels correspond well to those identified by ICMM (2012) of compliance, improvement, and learning. When it comes to formulating the goals then – that is, essentially answering why work environment management should be practices or implemented – this must be done with regards to the current organizational context. And as work with the work environment progresses, these goals need to be revised to suit the changed context.

Related to this, Harms-Ringdahl (2004) talked of three different work environment management strategies: a strategy of rules, knowledge dissemination, or economic incentive. In the 'rules strategy', the idea is that all would function as intended if only companies and employees would follow the rules. The 'knowledge dissemination strategy' assumes that if management and employees knew what is good (e.g. in terms of the working the environment), they would act correctly. Finally, the 'economic incentive strategy' builds on the idea that organizations and individuals will chose a certain action if it carries with it economic advantages. What should be clear from the book so far, however, is that while no strategy is entirely wrong per se, none of them are sufficient on their own: it is here that the holistic approach assumes importance. And the underlying assumptions are important as they will influence how, for example, goals are set or how they are pursued. In summing up this discussion, Harms-Ringdahl (2004) gave the following recommendations: avoid simplified explanations and solutions; investigate different incentives and interests and how they may be in conflict with each other; and view the work environment management as a negotiation and interaction between actors and their interests and values.

6.2.3   AN EXAMPLE: SYSTEMATIC WORK ENVIRONMENT MANAGEMENT AND BEYOND IN SWEDEN

We end this subsection with a brief account of systematic work environment management in Sweden – where it is mandated by law – together with an insight into how some Swedish mining companies view the strategy as well as other work environment management systems.

The Swedish Work Environment Act points out the responsibility of the employer and stipulates the basic demands on a good work environment. According to this act, the employer is always responsible to ensure that their own operations are handled in such a way that work-related accidents and ill-health are prevented and work is performed in a satisfactory work environment. The act emphasizes preventive actions and also cooperation between employers and employees. However, the cooperation does not diminish or abolish the employer's responsibility to carry out any measures necessary for the safety and health of the employees; the employer shall plan, steer, and control the work environment.

The Work Environment Act gives the framework for provisions issued by the Swedish Work Environment Authority. The Work Environment Authority's *Statute Book* contains more than 100 provisions, which are both very specific as well as more general. The provision of systematic work environment management (referred to as AFS 2001:1, or SAM after *Systematiskt arbetsmiljöarbete*, which literally translates to 'systematic work environment management') is the most important general work environment provision in Sweden. In a simplified way, it can be described as a compulsory continuous improvement process and a management system consisting of the following basic steps:

1. Investigate the present working conditions.
2. Carry out a risk assessment.
3. Deal with risks which can or need to be eliminated or reduced at once.
4. Formulate a work environment policy based on the present and the wanted/preferred situation.
5. Draw up an action plan for risks that need to be eliminated or reduced later on.
6. Draw up a clear allocation of work environment tasks, resources, and powers.
7. Make sure that the people who are allocated tasks also have the necessary competence to carry out the planned actions.
8. Follow-up of the effects of the performed actions.

This basic improvement process is not prescribed in detail and has to be adapted to the needs in the specific business or industry; extensive problems or businesses need more sophisticated systems. (Note the similarities to PDCA – plan, do, check, act – methodology and its principle of continuous improvements.) More details about the process are also stipulated in AFS 2001:1, such as that the top management must, at least once a year, control the efficiency of the improvement process in order to improve it. It also points out the importance of preventive actions, cooperation between management and labour, participation of all employees, and continuous improvement. The AFS 2001:1 has major similarities with the voluntary British standard OHSAS 18001 and OHSAS 18002 (OHSAS stand for 'Occupational Health and Safety Assessment System'), but AFS 2001:1 especially stresses the importance of cooperation between management and employees as well as their representatives. (The legislation in itself also corresponds well to the requirements of both the ILO Conventions on safety and health and EU Directives.) This cooperation is a strong and old tradition in Sweden where, workers' safety representatives (established already in 1912) hold an important position both in theory and in practice (they are well established during production work and the related work environment management but are less involved in initial planning phases). The AFS 2001:1 is also applicable to planned changes, where it specifies that when changes within the organization are planned, the employer must assess if these changes bring about risks of accident or ill-health that need to be rectified.

Lööw et al. (2017) investigated the improved accident frequency rate of two Swedish mining companies by interviewing company representatives with extensive knowledge on safety, work environment, technology, and organization. The study found that most respondents held that the AFS 2001:1 (and its precursor provision) had had an important and positive impact on the working environment in the mines. Some respondents phrased it as forcing the companies to adopt a systematic approach to these issues. For both companies, the systematic work environment management appeared to make up part of the foundation for subsequent efforts to improve the work environment: one of the companies invested in a safety culture programme, while the other focused on becoming certified in OHSAS 18001. Interestingly, following these efforts, the 'safety culture programme company' focused on getting certified for OHSAS 18001 as well, and the other company invested in a safety culture programme. That is, in both cases, the companies progressed from systematic work environment management to a work environment management system. The work environment management system appears to provide an additional structure to the work environment management, and many of the respondents argued that certification leads to higher safety. Others argued that the system was a way to sharpen the systematic work environment management.

As seen, there are several approaches to work environment management, and it is difficult to recommend any one starting point. What should be clear, however, is that work environment management is required, that it needs to be either systematic or part of a system (or both), and that it can function as a foundation for further work environment improvements and interventions.

6.3   IDENTIFYING, ASSESSING, AND TREATING RISKS

A central element in work environment management is to manage risks. The exact terminology used here varies significantly; Harms-Ringdahl (2013) identified the terms risk management, risk assessment, risk identification, risk analysis, risk evaluation, and risk treatment, as well as safety analysis. (For others still, talk is of work environment or workplace analysis.) Harms-Ringdahl (2013, p. 45) argued that 'It is difficult to distinguish between the terminologies, especially since there are differences between standards, and there are also changes over time'. In the probabilistic (or technical) tradition, however, he held that the relationship between some of the terms is the following. Risk management consists of the process of establishing the context, risk assessment, and risk treatment. Risk assessment in turn consists of risk identification, risk analysis, and risk evaluation. In this section, we generally make use of this process and terminology, but our focus is not on terms in themselves. Instead, we focus on their content that is similar between traditions and particular approaches (such as safety analysis and workplace analysis). That is, while terms might differ depending on a range of factors, the general procedure should be familiar throughout.

Risk is also a concept with several definitions, as we noted in Chapter 3. To reiterate, in daily speech, risk is often equated to chance or probability. Technically, however, risk is the product of the consequence of an accident and the probability that it will occur. It is easy, then, to think of risk in quantitative terms, but it also has qualitative features. Risk, and especially probability, is not always, and does not always have to be, quantifiable; it may be enough to conclude that an accident has a potential to occur and that this potential can be bigger or smaller depending on different factors without expressing this in numerical values. In fact, in many cases, much is gained through identifying which factors increase the probability or consequence of an accident, even without determining exactly by how much.

Managing risks through identification, assessment, and evaluation of risks have been used with good results in the mining industry (Joy 2004; Nelson 2011). However, we have chosen to include this section because the methodology used is at times simplistic and does not capture the complete picture. This is not to say that even simple risk analyses cannot be effective; they are valuable and have contributed to making mining safer and working environment better. But common and currently utilized methods tend to focus on a task, what can go wrong in that task, and the probability and consequence of that going wrong. The probability and consequence might be judged on a scale of one to three, where a score of one represents a trivial consequence or very rare occurrence, and a score of three is very serious or frequent. These scores are multiplied, resulting in a final score of one, two, three, four, six, or nine. The first two final scores, one and two, tend to mean that the risk is negligible; scores three, four, and six suggest that they have to be rectified; and a score of nine indicates that work cannot continue until the risk is treated. Table 6.1 gives such an example from a mining company's risk analysis. (This approach is similar to simplified versions of Job Safety Analysis or FMEA.) While this can be a good method for getting a preliminary estimation of risks, it suffers from some drawbacks:

- Risks have to be present in relation to a task. While risks outside of task can still be identified, they are not done so systematically.
- There is no systematic way of identifying the risks within the task. As it stands, the analyst's imagination dictates which risks are discovered.
- While the scales can be extended to allow for more detail, thresholds regarding interventions are more or less arbitrary.

We hold that safety and the work environment can be improved further by taking a more thorough approach to risk management. The purpose of this section is to outline such an approach to manage risks. As noted, there are many variants and the specifics of each variant might differ. Therefore, we focus first on the general procedure of the process; this is followed by an introduction to the principles supporting some specific methods.

6.3.1 THE GENERAL PROCEDURE

A continuous method of identifying, eliminating, and reducing risks in the workplace – throughout the whole lifecycle – is needed. Figure 6.1 illustrates the general procedure described in this section. The intention is to show the iterative nature of this process. Most illustrations of the process tend to indicate some iteration, but we want to highlight the more central nature of working iteratively here. It should also be noted that the activities demarcated here, which are based on Harms-Ringdahl (2013), may not be as clear in practice, where they tend to overlap. Additionally, some methods also progress differently. Again, however, the overall content should be present in most approaches. (Additionally, note again the similarity to the PDCA cycle.)

6.3.1.1 Establish the Context

Establishing the context is essentially about deciding what shall be investigated. In this stage, the study object (e.g. the whole workplace, a mine face, or a certain machine), definitions (e.g. what constitutes a risk or an accident), responsibilities, and so on are decided. The question 'Why is the analysis needed and how are its results going to be used?' (Harms-Ringdahl 2013) also has to be answered at this stage.

It is important to remember that at early stages of the work, such as in the design stage, a broad perspective is usually required. With more concrete projects, risks can be defined for smaller and more specific areas. For different mine planning projects, whether to use broader or narrower perspectives also depends on what mine phase the project regards. In the operating or exploitation phase of a mine, hazards at a task level can be studied. In the design phase, it is usually not known how the tasks will be carried out; thus, a broader level of perspective has to be used. Nonetheless, it is possible to know the job and its tasks in broader terms. For example, it is possible to know that a certain machine has to be used for a certain job. It is also possible to know what tasks are needed to operate the machine. The smaller the boundaries and the more details and knowledge there is about the system, process, or tasks, the more detailed the hazard identification will be.

TABLE 6.1

An Example of a Risk Analysis from a Mining Company

Operation/area	Risk	Consequence	Consequence	Probability	Risk	Countermeasure
Travel to workplace	Traffic accident with car	Injury to person	3	1	3	
Break room	Messy, risk for illness	Illness (injury to person), discomfort	2	2	4	5s, schedule cleaning
Break room	Mold	Health risk	2	1	2	Renovate or replace when mold occurs
Stope	Falling rock	Injury to person	3	1	3	Routines in place
Stope	Fire in transformer	Injury to person	3	1	3	Routines in place, use gas indicators, regularly scheduled measuring of radon, personal radon indicators
Stope	High doses of CO, NO, radon, and ammonia	Health risk, injury to person	3	2	6	
Stope	Open chute/ore passes, possible to fall in	Injury to person	3	1	3	Use shut-off chain
Stope	Collision between loader and other vehicles if other vehicles are in stope	Injury to person	3	1	3	Use gate or chain, contact operator of the loader if entering stope
Loader	Electrical current through machine due to grounding failure	Injury to person	3	1	3	Continuously check if grounding mechanisms are intact
Loader	Slipping risk due to oily machine	Injury to person	3	1	3	Rinse off machine, anti-slip strips on steps, daily supervision, 5s

(continued)

TABLE 6.1 (*Continued*)
An Example of a Risk Analysis from a Mining Company

Operation/ area	Risk	Consequence	Consequence	Probability	Risk	Countermeasure
Loader	Fire in machine	Injury to person, health risk due to smoke	3	1	3	Fire training is included in loading operation training, current escape paths are updated
Loader	Contact with wall	Injury to person	3	1	3	Use seat belt
Loader	Driving over big rocks	Injury to person	3	1	3	Awareness and attention when driving, maintenance of roads, use seat belt
Loader	Poor ergonomics in seat	Strain injury	2	2	4	Develop better seats, preventive body care
Oil container	Fire in oil container, confined by smoke	Injury to person, health risk due to smoke	3	1	3	Keep clean around it
Media	Faulty power central	Injury to person	2	2	4	Order repair or disassembly
Media	Electrical current in hanging cables	Injury to person	3	1	3	Order fix
Electric cable	Grounding error, fire hazard	No power to machine and risk of fire and injury to person	2	2	4	If grounding error alarm occurs, contact cable repair
Stope floor	Explosives left	Injury to person and machine	3	1	3	Visual control before loading that no undetonated explosives are present
Loading	Working alone	No direct help if something happens	1	3	3	Visual control before loading that no undetonated explosives are present

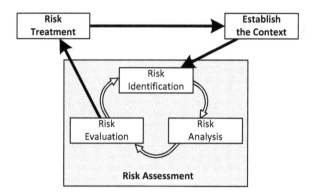

FIGURE 6.1   A general procedure for risk management. (Based on Harms-Ringdahl 2013.)

6.3.1.2   Risk Identification

The next step is risk identification. Here, the goal is to find and describe the risks. All situations in the investigated area that can cause accidents have to be identified. Damage or injury to a person, the environment, property, and equipment should be included (depending on the scope of the analysis). This is one of the most important steps as it is not possible to eliminate or control risks that are unknown. There are several available methods for systematically identifying risks; this step will depend much on the tools that are used.

At this stage, it is also critical to involve the relevant stakeholders. When identifying risks at a workplace, the employees must be included. But expert knowledge is also required. Findings from a study by Bahn (2013) is illustrative of this. She found that the length of underground experience did not determine the ability to identify risks, and that this ability varied extensively between groups. Those in supervisory positions identified few risks. So, there is a need to involve enough people that all risks can be identified. Additionally, it appears that methods that help in identifying risks could be of significant support here.

6.3.1.3   Risk Analysis

Risk analysis is about understanding the nature of the risk and determine its level. This means that in the risk analysis step, the causes and sources of risk are found. The consequence and probability of the risks are also determined (though this can vary depending on which methods is used; some methods for risk analysis do not estimate the consequence and probability). How this is specifically done depends on the chosen method. For an energy analysis, this could be calculating energies; in a deviation analysis, it could mean estimating the consequence of the deviations. The results from the analysis are used as input for the subsequent risk evaluation.

6.3.1.4   Risk Evaluation

After risks have been identified and analysed, they then have to be evaluated. In this step, it is decided if the risks are acceptable or not and how they should be dealt with. Risk evaluation includes decisions such as if a risk should be treated or not; in what order they should be treated; and if, as well as what, action should be taken. The method of evaluation will sometimes depend on the specific method employed, but there are two methods of evaluation that feature in many other methods.

6.3.1.4.1 Risk Matrices

One of the most popular ways to evaluate risks is to use a risk matrix (Harms-Ringdahl 2013). This method classifies a risk based on its likelihood of happening on one axis and its consequence on the other. This creates a grid or matrix (see Figure 6.2). From the risk analysis, the consequence of the risk and the likelihood of it happening is determined and can then be placed in the matrix based on these two attributes. This combination of consequence and probability results in a classification of the risk. In Figure 6.2 there are three classifications. Those risks classified as low (or similar) can be ignored, but those classified as high (or similar) have to be treated or solved immediately. Generally, for the highest classifications, the work, project, and so on cannot continue until these risks are treated. The risks in-between (in this case, those classified as 'medium') should be treated or investigated further, but they do not have the same priority; in many cases, it is a question of time span, that is, how soon the risk has to be dealt with (within a week, a month, etc.).

It is important that all levels are clearly defined when establishing the context. It is also common for the scale of the risk matrix (the likelihood and consequence) to vary in detail and definition. Likelihood can, for example, be graded as the risk occurring as often as once a week to as rare as once every 100 years. Definitions such as these can be useful when dealing with historical data and suchlike, but they are not as useful when dealing with conceptual systems and similar aspects. The probability can also be based on factors such as occurrence in the whole sector or industry. As such, the risks can be classified as 'Never being heard of in the industry/sector' at one end of the scale and 'Having happened more than once per year at the location' at the other end. These classifications may be more useful for conceptual evaluation of new machines, workplaces, etc., as it is possible to do a benchmark of sorts to get an estimation of the level of risk. However, when evaluating novel systems, most

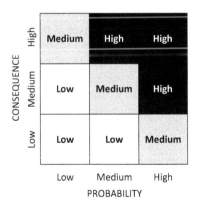

FIGURE 6.2 An example of a risk matrix.

risks risk being classified as 'Never being heard of in the industry/sector' because new technology can introduce new types of risk.

Consequence classifications vary in a similar way. The consequence of the risk could, for example, cause some discomfort but is otherwise harmless, injury but no sick leave, short sick leave, long sick leave, and permanent serious injury or death. A significant problem here is determining which consequence to consider: a risk can cause a number of different consequences. A simple misstep may result in no injury at all, or it can cause a sprain with a medium-term sick leave. At times, it may be necessary to list several consequences for a risk or have different consequences be assigned different probabilities. Another way of classifying consequences is determining if the risk can cause 'no injury or health effect' at one end of the scale to 'more than three fatalities' on the other end (but see Hopkins (2004) on these points). A classification such as this is more general but also vague; for example, there is no clear delimitation between a slight and minor injury. Consequences do not have to (and should not) only be limited to people. Often, consequences to assets, the environment and reputation, and so on, should also be considered.

Lately, the risk matrix and similar probability-based methods for assessment have been frequently criticized. Cox (2008) drew attention to the fact that, even though risk matrices are widely accepted and used, little research has been done towards validating their performance in improving risk management decisions. He continued by pointing out the limitations of risk matrices: at worst, when it comes to giving guidance in decisions, they are worse than random. He concluded that they should be used with caution and only with careful explanations of embedded judgements. Furthermore, demarcations can be arbitrary in nature, meaning that the requirement of intervention may be due to personal judgement. Despite this, risk matrices remain popular. Their application should, however, be characterized by carefulness and reflection regarding, among other factors, the estimation of the consequences and the probability as well as the consequence of misestimating these factors.

6.3.1.4.2   Direct Evaluation

An alternative to the risk matrix is to use direct evaluation. In a direct evaluation, the decision of doing something about a risk is based on criteria. This makes it more direct than a risk matrix. First, a list of criteria that a risk cannot violate is generated. Then, each risk is evaluated to determine if they violate one or more of the criteria. If this is the case, this is ground for doing something about the risk. Criteria be that a risk should not break company policy or any authority directive, or it can stipulate that a risk cannot have a great consequence or likelihood of occurring. When a risk violates one of the criteria, something has to be done about it; when a risk violates more than one criterion, it means that it has a higher priority. Harms-Ringdahl (2013, p. 70) gave the following criteria as examples of what risks can be evaluated against:

- 'Is a breach of authority directives'
- 'Deviates from the company's policies and/or rules'
- 'Has big consequences and/or high probabilities'
- 'Deviates from good praxis'

- 'Involves many uncertainties (i.e. there is too little knowledge about the risk)'
- 'Can be easily eliminated or reduced by an obvious solution'
- 'Can affect a system with low tolerance for error and/or faults'

Other criteria can be more specific, for example, a consequence of a risk cannot cause a fatality, or is not accepted by a stakeholder. The advantage of the direct evaluation method is that it is less ambiguous, and that it is relatively clear why a risk has to be treated.

6.3.1.5 Risk Treatment

Risk treatment is the next step: having decided that something has to be done about a risk, it has to be treated. In this step, the goal is to find ways to reduce or eliminate the risk. Reducing risks can be done through reducing its likelihood or its consequence (or both) and can be done in several ways; what is important is that it is done systematically. A strong recommendation is to use the hierarchy of control, which is a popular tool for risk treatment. There are different conceptions of the hierarchy of control, with varying numbers of levels, terminology, and so on. Most, however, are related to the work of Haddon (e.g. Haddon 1995). Haddon noted that for something to cause an injury, the energy that affects an object must surpass a certain threshold and proposed ten countermeasure strategies for injury prevention with reference to these energies (see Table 6.2).

In the hierarchy of control (whatever the conceptualization), the first attempt to treat a risk is made at the top level. If the first attempt is not possible, an attempt is made at the next level. The procedure is repeated until a suitable solution is found; if the risk cannot be treated at the current level, the next level has to be attempted. Progressing in this manner can also have economic benefits. A study by the European

---

TABLE 6.2

Ten Strategies to Prevent Damage from Energies

Strategy	Description
First strategy	Prevent energy accumulation in the first place
Second strategy	Reduce the size of the energy
Third strategy	Prevent the release of the energy
Fourth strategy	Reduce the rate of energy release
Fifth strategy	Separate energy and object
Sixth strategy	Separate energy and object through barriers
Seventh strategy	Modify the contact surface, basic structure, and so on (e.g. round sharp edges)
Eight strategy	Strengthen the object
Ninth strategy	Rapid response to damage
Tenth strategy	Return to stable state

Source: Based on Haddon (1995).

Commission (2011) investigated the costs and benefits of prevention measures. The highest benefit–cost ratio was generally found for measures aimed at substitution or avoidance. The lowest ratio was found for measures such as training and personal protective equipment. This indicates that the most effective measures, such as substitution and avoidance, are also more cost-effective. Furthermore, these measures are easier to implement in early project stages, before machines have been bought, layouts constructed, and so on.

There are some treatments or interventions that do not fall under the hierarchy of control. One such example is when there is not sufficient information available to make a fair decision: the risk 'treatment' can then be to gather more information (Harms-Ringdahl 2013). It is also important to note that all treatments may not be possible to implement immediately. As recognized in many work environment management concepts, a plan of action can be established that details how, when, and by whom a risk will be treated. It is also recognized that the measures need to be followed up to ensure that they function as intended.

Below we describe the hierarchy of control as consisting of three levels (see Figure 6.3). These are based on Safe Work Australia (2011).

6.3.1.5.1 Level 1

At the first level, risks are eliminated completely. This is always the most effective, but not always an efficient, solution. An example of this level would be eliminating the risk of falling by performing the work at ground level or using automation and remote control to eliminate manual or mechanized underground work. It is not always efficient because it could sometimes risk removing the value-creating activity itself (if no activity is performed, there is no risk).

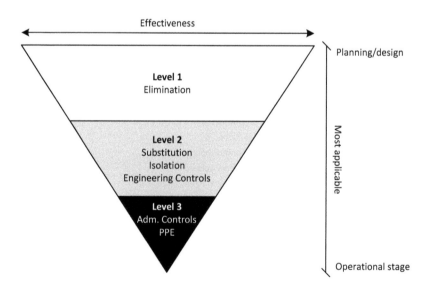

FIGURE 6.3   A conceptualization of the hierarchy of control. (Based on Safe Work Australia 2011.)

6.3.1.5.2   Level 2

At the second level, risks are minimized and reduced. There are three different strategies employed at this level; all aim at reducing the consequence or the like-lihood of the risk. First, an attempt to minimize the risk by substitution is made, that is, by replacing the (source of the) risk with something safer, for example, using water-based paints instead of solvent-based paints or, to reduce dust emissions, using down-the-hole drilling with water hydraulics instead of pneumatics.

The second strategy is to isolate the risk from people (or other vulnerable objects) or people from the risk (though isolating the risk from people is preferable). This means physically separating (or isolating) risks and people by distance, time, or bar-riers. Installing guardrails around holes in the floor is an example of this strategy; mining machines with safety cabins and good climate control is another. Planning noisy or otherwise disruptive activities at times of the day when there are few miners present is also an example.

With the third strategy, engineering controls are used. Risks are reduced by changing the workplace or work organization; for example, using trolleys to move heavy loads instead of carrying it manually or decreasing work hours to reduce fatigue. Note that these do not aim to change the behaviour of people or putting the responsibility for safety on operators.

6.3.1.5.3   Level 3

Personal protective equipment (PPE) and administrative control are found in the third level. These should only be used as a last resort. Administrative control means using work methods or procedures designed to, for example, minimize the exposure to a risk. This can mean using warning signs, worker training, or guidelines for safe machine operation. A common example of administrative control is the procedure for safe handling of explosives.

The alternative is to use PPE, which includes hard hats, gloves, and protective eyewear. PPE can be efficient, but it is important to remember that PPE only limits the harmful effects of a risk *if* they are utilized in the first place, and then also cor-rectly used; in mining, it is common for PPE not to be used or misused (McBride 2004).

6.3.2   CHOOSING A METHOD

Beyond the general procedure, there are many useful tools and methods. Each has its specific application area and, usually, its own dedicated literature (see Harms-Ringdahl 2013). Choosing a proper analysis method is critical. This choice is com-plicated by the plurality of options and their appropriateness for the intended stage, application area, and so on. Harms-Ringdahl (2013) listed the following variables for consideration when choosing a method: the objective of the study; availability of resources (people, time, and budget); what the business is doing; regulatory and contractual requirements; in what stage of the lifecycle the business is in; and what information is available. It is also a good idea to use more than one method to cover more parts of the risk spectrum. Harms-Ringdahl (2013) reported on studies and experiences that show how energy analysis, deviation analysis, and function analysis identified roughly the same number of hazards; but only 5 per cent were generated by all three methods. As different safety analysis tools provide different outcomes, there is a need to use a number of different tools to cover all relevant areas. The study also indicated that it is important to be familiar with a number of tools in order to choose and use the one that best fits the need of the organization at any particular time and situation. Shooks et al. (2014) compiled several available tools alongside their suitable application stages (see Table 6.3).

TABLE 6.3
Available Methods and Their Application

Project Life Cycle Stages	Safety Analysis Technique	Purpose
Conceptual	Preliminary hazard analysis (PHA) – Initial	Identify hazardous situations and events within concept design
		Prioritize hazards for further analysis
		Identify potential significant project health and safety risks
		Develop recommended actions
	Checklist	Determine compliance to standards and legal framework
		Check that everything has been covered
Prefeasibility	PHA – Review	Review for design changes from concept design or develop new PHA
	What if?	Identify and rank according to risk how major unit operation will be affected by deviations from normal operations and behaviour
		Provide a basis for a risk register
	Checklist	As above
Feasibility	hazard and operability study (HAZOP)	Assess process-control design hazards/risks
		Define possible deviations from the expected or intended performance
		Generate consequences, controls, and control actions
		Identify hazards introduced through human actions
	Failure mode and effect analysis (FMEA)	Identify potential failure of various parts of a system
		Estimate the effect of the failure and how to avoid and mitigate the effects of the failures on the system
		Assist in selecting design alternatives with high dependability
		Identify human error modes and effects

(Continued)

TABLE 6.3 (*Continued*)
Available Methods and Their Application

Project Life Cycle Stages	Safety Analysis Technique	Purpose
	Fault tree analysis (FTA)	Identify and analyse factors that can contribute to a specified undesired event
		Illustrate factors and their logical relationship to the undesired event
Execution and Construction	HAZOP	Review feasibility HAZOP
	Construction risk assessment workshop (CRAW)	CRAW per work package/area/construction type
		Identify risks and controls during execution of project (including contractors)
	Job safety and environmental analysis (JSEA)	Identify hazards and controls for tasks
		Assess the risks of tasks
		To be used in the development of work procedures
	Stop, look, analyse, manage (SLAM) or Take 5	Identify hazard and control methods for tasks
		To be used by individuals
	Authority to work (ATW) process	Ensure all hazards and risks have been assessed and taken into account prior to starting a specific activity
Commissioning and Ramp-Up	Checklist	As above
	FMEA	As above
	CRAW	As above
	JSEA	As above
	SLAM or Take 5	As above
Operation	HAZOP	As above
	SLAM or Take 5	As above
	JSA	As above
	Checklist	As above
	FTA	As above
	JSEA	As above

Source: Shooks et al. (2014).

6.3.2.1 Two Recommended Methods: Energy and Deviation Analysis

While most methods should be chosen to fit the current situation and context, there are two that are applicable to most situations. Johansson et al. (2010) argued that classical tools for identifying occupational risks in production environments, such as safety rounds and incident and accident reporting, are less appropriate for identifying and assessing risks in future and planned work environments. Instead, they held that proactive methods, such as preventive deviation and energy analyses, are required. While we do not cover these methods in detail, we provide an account of their general principles to give a sense of how they can be used.

In deviation analysis, a deviation is defined as an event or condition that deviates from the intended or normal. It takes into account the entire system: humans, technology, and organization. The purpose of deviation analysis is to predict and to prevent deviations that can cause damage (Johansson et al. 2010). A deviation can, for example, be forgetting a step in a procedure, making a wrong turn, or pressing a wrong button (cf. human errors; see Section 3.3.2). The deviation analysis does not focus on accidents but on identifying every possibility of performing a task in a deviating manner (this is usually done with the help of a checklist). All deviations may not lead to an accident or may only lead to an accident if combined with another deviation. This gives priority for the treatment of the identified risks: deviations that would lead directly to an accident need to be treated first, while those requiring several deviations are of lower priority. Deviations of equal priority can be prioritized according to probability and consequence.

Johansson et al. (2010) described energy analysis as consisting of three main components: energy that can cause damage, targets that may be harmed, and barriers to energy. The energies usually considered are gravity, height (including static load), linear motion, rotary motion, stored pressure, electrical energy, heating and cooling, fire and explosion, chemical effects, radiation, and miscellaneous (human movement, sharp edges, and points). The energy analysis can consider tasks as well as physically demarcated areas. For each area or task, the analyst checks if any of the energies are present as well as identifies any potential barrier to the energy. While the assessment can still judge the risk by estimating probability and consequence, it is also possible to have thresholds for the energies involved. For example, it could be decided that no energy of more than x N m should have the possibility to reach and interact with an operator. This means that any energies that surpass this threshold have to be reduced to below x N m with barriers or ensuring the operator never interacts with the energy (e.g. by moving the operator).

An advantage with methods such as these is that they allow the evaluation of conceptual systems (hence the focus on *preventive*). While the specifics may depend on actual implementation, large parts of a concept can be evaluated before that. For example, calculating energies for new machines is quite simple, as is investigating what happens if a remote control system is operated in the wrong order. Recently, we have begun teaching these methods to mining and civil engineering students. They were able to use the two methods to analyse conceptual systems to identify and suggest measures for treating risks.

6.4 INDICATORS

Shaping and monitoring the work environment and work organization in mining is complex; relations between causes and effects on health and safety are often diffuse, and causes can interact in several different ways. Additionally, many effects are revealed only after a long exposure time. Therefore, it is necessary to use a number of indicators of different types to shape and monitor the work environment and the work organization.

Indicators can be categorized as leading indicators (also called 'activity' and 'before-the-fact' indicators) or lagging indicators (also called 'outcome' and 'after-the-fact' indicators). Leading indicators measure the direct precursors to harm or damage; they provide 'warning signals' before an unwanted event or deviation occurs that could lead to an incident or an accident. This can prevent serious loss or damage to people, environment, or property. If leading indicators that function are available, they are very valuable. However, leading indicators are often difficult to implement and can usually not be used for external benchmarking (ICMM 2012).

Lagging indicators, on the other hand, measure final outcomes; they identify the hazard once it has manifested. Lagging indicators are most widely reported in the mining industry (Ekevall et al. 2008). These indicators are generally easy to collect and measure, and due to regulatory reporting requirements, they tend to be mandated (ICMM 2012). Three internationally common indicators used are lost time injury frequency rates (LTIFR), disabling injury severity rates (DISR), and fatal injury frequency rates (FIFR). Other classic, and in most cases well-functioning, indicators were identified by Johansson and Johansson (2008) as the number and rates of fatalities, accident numbers and rates, numbers and rates of occupational diseases, sick leave statistics, and the number and rates of and reasons for early retirement pension. Gradually, however, focus and interest has shifted towards the leading indicator as they focus on the future and are proactive (Ekevall et al. 2008). Nevertheless, to effectively monitor and manage the work environment, both types of indicators are required.

6.4.1 REACTIVE AND LAGGING INDICATORS

The LTIFR is the number of lost time claims per million hours worked. Lost time claims indicate that the worker has had time off due to the injury and can thus be regarded as a more serious, non-trivial injury. The rate makes comparisons possible between different business sectors, different time periods, and different companies. It is possibly the most used health and safety index worldwide, and considerable energy and focus is spent on keeping the LTIFR low. However, this index has a blind spot regarding fatalities: a fatality and a simple skin cut injury, where a person has a whole day off, give the same statistical result regarding the LTIFR. Therefore, this index is criticized for trivializing personal damage and directing preventive efforts on minor risks that do not have the potential to result in serious injuries; it is said that the important emphasis on serious personal damage is lost with too much focus on LTIFR. Preventing minor accidents will not automatically prevent major accidents; preventing fatalities and severe injuries is much more complex and difficult than bringing down the LTIFR. Correctly managed, however, the LTIFR is a valuable index and tool, but it needs other indexes as a complement. For example, when comparing between countries, or even between companies, there can be differences in reporting systems, social insurance systems, business practices, culture, etc., which can make such comparisons misleading. For this type of comparison, the FIFR is a more certain and comparable measure over time. In countries with a good safety record, this rate varies a lot over the years since the number of fatalities is low and a few fatalities engender large changes in the frequency rate. Additionally, where safety records are high because of high levels of mechanization and automation, single fatalities engender an even higher 'peak' in the FIFR. In countries with a low safety record, this rate can be very high. For example, on a continent level, the FIFR is more than twice as high in Africa compared to Europe and the Americas. These rates in turn are half of those in Asia. Oceania has more than half the FIFR of Europe and the Americas (ICMM n.d.).

For mining in particular, an important aspect is the extensive utilization of contractors. Statistics for this group is collected and reported to an increasing extent, but there are some issues to be aware of. A common problem seems to be that many contractors do not report all near-misses and accidents. This is alarming since present statistics indicate that contractors' LITFR have been two to four times higher than the rates for the mining companies' own personnel. It is obvious that the situation for the contractors must also be regarded when judging how safe mining activities are. (In the mid-1990s, Blank et al. (1995) found that the official statistics for the Swedish mining branch did not reflect the real risks. Their research also indicated that contractors had more frequent and severe accidents, and that they performed other tasks under other conditions than the mining company employees.)

More recently, some new indicators have been used in Sweden. For example, different employee satisfaction indexes have become more common among Swedish enterprises. One version of this index is an aggregate measure of employee satisfaction with 11 central areas: participation, information, wages, work satisfaction, physical work environment, security and comfort, work load, leadership, trust, goals and follow-up, and competence development. The index can be customized to different types of industry. Health promotion activities and different lifestyle indexes are also becoming more common. These can include the percentage of employees with no sick leave during a year, which is measured in many companies. Additionally, investigations have shown that obesity is an increasing problem in Sweden and among miners. About 20 per cent of the Swedish miners have a body mass index classified as obesity with significant consequences for the miner's health. This type of measure could also serve as a useful indicator.

It is also possible to look beyond factors that are directly concerned with employees. For example, monitoring occurring fires and fire incidents is extremely important in mining due to the potential fatal consequences of fires. These statistics are often presented as number of fires per year, categorized by severity and cause. Similarly, statistics regarding production stops, machine breakdown, and so on can be tracked. While they may not directly affect safety or factors relating to attractiveness, they do so indirectly. Looking at causes may be particularly appropriate as they may reveal errors that occur due to factors such as fatigue or boredom (see Section 6.4.3).

In sum, using reactive indicators is useful, but they also have to be viewed critically and complemented with other measures to give a fair picture. It should also be remembered that they should not be goals in and of themselves. For example, keeping a rate below a certain threshold should only be a goal to the extent that it represents a good work environment; as noted, maintaining a low LTIFR by only addressing lighter, easy-to-fix injuries is a misuse of the indicator. And as a final note, Frick and Wren (2000) remarked on the reactive indicator's connection to work environment management. They argued that work environment management systems 'that are evaluated using [lost-time injuries] and other measures of employee behaviour mostly explain injuries as commonly caused by the behaviour of the victim, rather than by hazards in the environment. This usually results in an emphasis upon "safe people" rather than "safe places"' (Frick and Wren 2000, pp. 26–27). While safety and workplace attractiveness require attention on both behaviour and external factors (such as organization and technology), focus should be on the latter.

6.4.2 PROACTIVE OR LEADING INDICATORS

The use of proactive indicators is not yet common in the mining industry but is receiving increased attention. If focused on relevant factors, these indicators can make the management system for work environment issues more efficient. ICMM (2012) argued that proactive indicators should be easy to understand for responsible parties, have a clear effect (i.e. the indicators corresponds to a desired effect), be designed so that it drives continuous improvement work, and fit with other strategic indicators. ICMM also noted that proactive indicators are less static than reactive indicators; proactive indicators evolve together with the organization. Thus, they are chiefly internal and company-specific, which makes them less suitable for external benchmarking (ICMM 2012).

ICMM (2012) also argued that the ability for mining companies to use proactive indicators is dependent on their level of maturity. They identified three levels of maturity. The first level, compliance, is concerned with indicators that stem from legal demands, that is, the most basic demands regarding control and monitoring of the workplace. At the second level, improvement, the indicators are connected to preventive measures within a work environment management system. At the third and final level, learning, the indicators are developed from the specific context, such as risks specific to a certain company. The three levels are exemplified with reference to indicators as follows (ICMM 2012):

- Level 1 (compliance)
  - There is a work environment policy.
  - The percentage of employees who have completed safety training.
  - How often work environment questions are included in company communication.
- Level 2 (improvement)
  - The work environment policy is communicated to the whole organization.
  - There are surveys for understanding how employees perceive how management prioritize work environment questions.
  - The percentage of planned versus performed risk analyses.
- Level 3 (learning)
  - Percentage of work tasks that have been risk-analysed.
  - The number of work environment improvements shared between workplaces.

- There are surveys for understanding how employees perceive work environment information and training.

Because of the nature of proactive indicators, it is difficult to give any definite recommendations regarding them. As with reactive indicators, however, to give a more complete picture, they need to be used in a combination with other factors. These other measures can be both proactive and reactive; it is the appropriate combination and measurement that is important.

6.4.3 PERSONAL AND PROCESS SAFETY INDICATORS

There is a third category of indicators that do not directly distinguish between being reactive or proactive. These were referred to by Hopkins (2009) as personal and process safety indicators. The two types are described in Table 6.4. The personal safety indicator is concerned with the individual and its exposure to, for example, accidents. The process safety indicator, on the other hand, is concerned with the effectiveness of the control that the risk-control system relies on. One advantage with this type of indicator is that it focuses less on the individual and more on the process or system itself. Using the example of number of fires (mentioned in Section 6.4.1), the argument is that if there were several fires, even if no one was hurt, this is indicative of the failure of crucial controls; the only difference between a fire with no fatality and a fire with fatalities may come down to personnel happening to be in the right place at the right time.

6.5 ON WORK ENVIRONMENT ISSUES IN DESIGN AND CONSIDERING THE WHOLE LIFE-CYCLE

The Swedish provision on systematic work environment management already stipulates that its process is applicable also when planning, that is, risk analyses and so on

TABLE 6.4
Personal and Process Safety Indicators

Indicator	Explanation	Examples
Personal safety	Hazards affecting individuals	Falls
	Have little to do with the processing activity of the plant	Trips
		Crushing injuries
		Vehicle accidents
Process safety	Hazards arising from the processing activity in which the plant may be engaged	Escape of toxic substances
	Usually damages the plant or have the potential to damage plant	Release of flammable material
	Potential to cause multiple fatalities	

Source: Based on Hopkins (2009).

are to be conducted even for planned change. This perspective can be extended even further to argue that the work environment must be managed through all the project stages: beginning with planning and ending with decommission. 'Incorporating human factors early and often in the equipment design process and life cycle is … central … One way to broadly achieve this is through *safety in design* (sometimes known as *safe design* or *prevention through design*)', as Horberry et al. (2011, p. 18, emphasis in original) put it. Much of what can be said regarding human factors also holds true for safety, ergonomics, and workplace attractiveness. As such, for this question in general, we refer to work Horberry et al. (2011) and related work. However, as we are concerned with the design of workplaces, there are a few aspects to which we want to bring attention.

In mining, as opposed to many other industries (though the construction industry is similar to the mining industry in this regard), workplaces are continuously constructed. On the one hand, this can be when new main levels are constructed in underground mining. This is usually a large undertaking that involves activities closely related to the construction industry (in fact, contractors from the construction industry are often used). In this case, then, safety design is about incorporating safety considerations at the planning and design phase of a mining project. Gambatese et al. (2008) found that several safety problems at construction workplaces could be connected to construction and design. One problem has been that the norms and standards that exist around how constructions and components should be designed have mainly revolved around the final user's health and safety and have thus neglected the work environment aspects during the building stage (Gambatese 2000). Professionals involved in the early planning and design phases have to take a greater responsibility for health and safety on the construction sites. And there are examples of health and safety questions being taken into consideration in these early stages (see Cameron and Hare 2008).

On the other hand, there are the more frequently constructed workplaces, for example, rooms or galleries in cut-and-fill mining or production drifts. One point of note here is that when we talk of attractive workplaces, these considerations must also include development work. And this type of work is probably harder to make it attractive because it is not as readily automated and moved to control rooms. Additionally, employing contractors is common in these stages, which can complicate the matter further. Another aspect is that time and effort must be spent to ensure that the health and safety aspects of even temporary structures, installations, and so on are adequate. At this stage, though, we approach questions of the wider context and how to consciously plan and design for safe, ergonomic, and attractive workplaces in a holistic perspective; this is the focus of the next chapter.

7 An Iterative Design and Planning Process for Designing Ergonomic, Safe, and Attractive Mining Workplaces

7.1 PLANNING AND DESIGNING MINING WORKPLACES

The planning and design of mining workplaces is closely related to the planning and design of the mine; it depends on decisions made by mine planners, mining engineers, and other similar professions. Mining operations involve equipment and infrastructure that is expensive to upgrade, high capital costs, and long lifespans. This means that mining operations 'can potentially remain captive to technology decisions made many years previously' (Bartos 2007, p. 153). With large underground mines that take 5–10 years to develop (Nelson 2011), newly started mining operations can be using old technology by the time they are brought on stream or at least, be affected by outdated decisions. This also makes it difficult to change decisions. Moreover, Bartos (2007) argued that mining companies are adaptors rather than innovators. This means they rely on third parties such as equipment providers for new technology (Hood 2004). Simpson et al. (2009) noted that there is a mismatch between what mining companies and equipment providers consider important and in knowledge of relevant issues. Adding to this, new technology can take 7–10 years to develop (Bartos 2007); by the time the technology is implemented, the problem that the technology was meant to solve may therefore no longer be relevant, of a different nature, or even solved through other means.

This is to say that even though the activities of mine planners and mining engineers significantly affect the work environment, they often do not have extensive knowledge about the subject. If planners, designers, and engineers do not fully or successfully consider work environmental factors, they can introduce negative, long-lasting effects into the mining environment. Attending to these issues once the workplace is in operation is expensive and opportunities to do so are limited. It is troubling that work with environment and safety issues often appear to be left unattended in the early stages of mine planning and development projects, especially since these issues are best managed systematically and addressed in the very first steps of (and then, continuously throughout) a project: the most efficient ways to achieve a good work environment is through proactive planning instead of reactive corrective actions. It is also the best way to reduce the costs of achieving a good work environment. While the slogan 'Safety First' has been established in the mining industry for some time now, it is at times still just a slogan: safety does not always come first, especially if the business has financial problems. And decisions that favour short-term gains, common during financial hardship, introduce subsequent losses later through cumbersome designs that , for example, create stress that in turn decrease performance.

Many mining engineering handbooks cover health, safety, and more general work environmental issues but tend to focus on these issues in the production stages, have a technical focus, and do not cover them in wider perspective. In contrast, this chapter is about how to work with issues of ergonomics, safety, and attractive workplaces in all stages and types of mining-related projects; for example, pre-studies for new main levels, development of new technology, or smaller everyday changes to the workplace. The approach that we propose is an iterative design and planning process.

7.2   THE NEED FOR AN ITERATIVE DESIGN PROCESS AND ITS POTENTIAL BENEFITS

It is not likely that mining will develop in a revolutionary way; evolutionary change, where incremental changes and improvements eventually lead to the visions of modern mining, is a more likely scenario. This means that much can be learned from history as well as the present state. For example, thorough evaluations of present and historic designs have systematically been used by LKAB in the design of their latest main level at 1,365 m below ground. This evaluation has been important because the time between the first conceptual designs and the final plans has stretched over 12 years and involved a large number of planners. Any project that will eventually affect the work environment (which is most projects) must therefore take into consideration and assess the historic, current, and future state (e.g. using risk analyses or other methods for analysis).

It is, of course, easier to assess and analyse present or historical issues than future issues. Even more so if the future holds large changes in technology, work organization, and so on. But even establishing which issues are relevant to consider in the present can be difficult. At times, which issues are relevant may become obvious only after a certain time in the project. In linear development processes, which are common in the mining and other industries, this is problematic as such processes do not easily allow for 'completed' stages or activities to be revisited. For these and other reasons (see below), projects can benefit from an iterative process. To some extent, all planning or design processes require rework and several iterations. But today, most of this rework and the numerous iterations are reactive and unsystematic: ad hoc solutions are found to problems whenever they emerge during design and planning activities. It is possible, and necessary, to have a more deliberate, proactive, and systematic approach.

Another important question is who is involved in the process. Usually, at least early in projects, engineers and those with a mainly technical focus tend to be involved. Even if these persons are knowledgeable about work environmental issues (which may not be very common), it is likely that they do not share the same view or perspective as operators, maintainers, and other personnel. Furthermore, all these different *stakeholders* will have different expectations on the outcome of the project, their own goals, etc. On the one hand, this means that relevant issues might be overlooked at a stage where they are best solved (if they can be solved at all in later stages). On the other hand, this will influence the goals of the project. Frick and Wren (2000) argued, for example, that if the work environment problems of the current workplace are defined as careless workers causing accidents (and neglecting to report them), then the object becomes the worker and the goal to control them. If instead the aim becomes 'to satisfy workers' desires for safe and sound work, the workers become actors, who are able to influence the integrated management of [the work environment]' (Frick and Wren 2000, p. 39).

When dealing with challenges of providing attractive workplaces, there is an additional aspect. As noted in the previous chapter, Harms-Ringdahl (2004) identified three actors: the individuals/employees, the organization/employer, and the society/institutions. Here, we conceptualize the last actor as the *surrounding* society because, in terms of attractiveness, it is imperative that the surrounding society has a bearing on the development on mining workplaces; the problem to be alleviated is essentially one of incompatibility between the individuals and the organization on the one hand, and the surrounding society on the other. That is, individuals of the surrounding society are not interested in being an employee of a mining organization.

In short, this is the outline of the motivation behind using an iterative design process, as well as of the method itself. The next section looks at some examples from outside mining to further discuss the need of this type of approach and its potential advantages.

7.2.1 THE USE OF DESIGN PROCESSES IN OTHER INDUSTRIES AND SECTORS

Ranhagen (2004) is central to the approach that we detail later. For this reason, we give a quite detailed account of what we consider the most important parts of his view on how to plan for a good working environment and sustainable working life. He held that these questions are questions of community planning: a good working environment cannot be achieved by only looking 'inwards'. This notion appears to be rooted in his background in architecture. Consequently, he emphasised that an organization's buildings and environments have become part of an organization's brand. To this end, they are designed in such a way that they support the development of identities and certain ways of working. The notion aligns well with the challenge of designing attractive workplaces. This entails both attracting those outside the company and retaining its current employees. It is a question of, for example, using the constructed environment to communicate values or facilitate certain ways of working that, on the one hand, are attractive to the current workforce and, on the other hand, are also attractive to a new workforce. That is, when designing for a new workforce, it is not enough for mining companies to look inwards; they must also take into consideration the wider society. Here, an important recognition is that it is not only about the 'what' and the end-results:

> Planning for a good working environment is not just about how to design the city, district, house, locale, or individual workplace to support a sustainable working life. Just as important is how the planning and projecting process is organized and designed. A carefully designed *planning process* has been shown to be as important as the *physical* design of the ... workplace, district, house, locale, or workplace.
>
> (Ranhagen 2004, pp. 289–290, our translation and emphasis.)

The 'design' of the planning process we cover further on; for now, it is important to note the other aspect of the process: participation. Ranhagen (2004) argued that the active participation and influence of the users are crucial. The notion of participative approaches is widely recognized and has been for some time (cf. user- or human-centric design). We emphasise, however, that Ranhagen (2004) takes a necessarily wider approach to the issues at hand. We are of the view that user- or human centric design, at least when it comes to workplace design, tends to overlook needs for a new workforce. Additionally, we argue that Ranhagen (2004) has a different focus on the implementation part of projects, as well as the connections between participation and aspects not directly related to the function of the 'product'. For example, he declared that active participation leads to greater engagement and taking of responsibility in the continuous adaption, development, and improvement of the working environment (i.e. even after the project is finished). But for participation to be effective, it must be initiated early, during strategic planning, and then continuously during the different stages of the project (Ranhagen 2004). To illustrate, Ranhagen (2004) introduced a rough rule of thumb that stipulates that what costs 100 dollars to change in a blueprint, costs 1,000 dollars to change at installation, and at least 10,000 dollars to modify after construction has started. Because user participation must actively influence designs, the users must be engaged at a stage where designs can still be changed.

Beyond this, Ranhagen (2004) listed several advantages of being attentive in and to the early project stages compared to neglecting these stages (see Table 7.1). He summarized this notion as the 'long–short' outcome as opposed to the 'short–long' outcome. In the 'long–short' outcome, the planning and execution of the project is given ample time. While this means that *projects* take longer time to complete, Ranhagen (2004) argued that it shortens the implementation time. It also has other benefits:

> It is important to maintain the ability to influence and the freedom of action as long as possible in process so that one can utilize the knowledge and insight the project participants acquire during the process. In the end, the solid preparatory work means that implementation will be less complicated and faster.
>
> (Ranhagen 2004, p. 291, our translation.)

By comparison, the 'short–long' outcome means that the project itself might be rapidly completed but would instead incur costs in the form of extensive rework and a longer implementation time.

The last part we want to highlight is the need for long-term thinking. Ranhagen (2004) argued that projects are rarely executed with a 'bigger-picture' perspective

TABLE 7.1

Being Attentive to the Early Project Stages Versus Trying to Rapidly Pass Through Them

Focusing on Early Project Stages (Long–Short)	Focusing on Quickly Complete the Project (Short–Long)
Smoother, cheaper, and more problem-free project execution	Total completion time prolonged
Lower operating costs and better work environmental standard during operating stages	Solutions that are time-dependent and difficult to modify
Solutions with greater flexibility that facilitates continuous modifications according to new demands that surface during operations	Increased life-cycle costs and decreased ability to modify the work environment as new demands become relevant

Source: Based on Ranhagen (2004).

in mind: projects are treated as isolated activities. For example, an expansion may be planned and completed with only the intention to satisfy current needs. This may constrain future expansion or changes due to how structures and infrastructure are constructed. Therefore, long-term visions or plans that outline the future state are needed. These plans recognize that each project constitutes single steps towards the end goal (cf. the 'evolutionary change' mentioned earlier). In Chapter 4, the notion was introduced that, to realize fully automated mining, mines need to be designed from the beginning with goal of being fully automated. This constitutes one type of long-term vision that outlines a future state; changes made today must align with, or at least not hinder, this vision.

In the process industry, there are similar notions of design. Most notably, Broms and Lindahl (2005) provided guidelines for achieving effective (including attractive) work in the process industry. Building on extensive experience, they found the participation of engaged and responsible employees are central to designing modern, effective, and attractive control rooms (or control centres; cf. Section 4.3). This builds on the idea that change in the process industry also includes technological change (this is a similarity that the process industry shares with the mining industry). As previous chapters have shown, any change should also offer significant opportunities for work organizational change. Broms and Lindahl (2005) argued that it is through appropriate work organization that the full potential of new technology is reached and efficient and attractive work can be created. Again, because it is the employees who are affected by and in a sense create the work organization, their participation is central to the design of new technology; the technology must facilitate the desired work organization of the employees.

Similar approaches have also been utilized in research projects. One example is the research project the *Future Factory: A Vision of the Future's Effective and Attractive Industrial Work Environment* (see Wikberg-Nilsson et al. 2011; Wikberg-Nilsson et al. 2009; Wikberg-Nilsson and Johansson 2010). The project was an interactive cooperation between different actors in the industry, such as managers, employees, youths, students, and representatives from unions and employer organizations. It aimed to generate a vision for effective, attractive, and sustainable industrial work environments. The project built on the idea of concept cars in the automotive industry. Concept cars are a way of marketing and testing new ideas; it is rare that these cars are actually produced. The purpose is instead to draw attention to the brand. The thinking behind the *Future Factory* was 'to develop a concept *factory* in order to draw attention to … industrial organizations and their design' (Wikberg-Nilsson and Johansson 2010, p. 2, our emphasis). We especially want to highlight two aspects here. First, the design teams involved in the project consisted only of women. The reasoning for this was that Swedish industry, in this case, is not able to attract women. As Swedish industry is male-dominated, it is also likely that it is typically men that are involved in traditional design processes. Hence, solutions are based on men's ideals and visions. To investigate the gap between ideals and visions between men and women, it was deemed necessary to constitute design teams that only consisted of women. This reasoning is also relevant for the design of attractive mining workplaces. The second aspect regards the project's use of participatory workshops in which the participants worked with, first, critiquing the present state, second, creating a future vision, and third, how to move from the present state to the desired future state. Wikberg-Nilsson and Johansson (2010) argued that the collaborative approach itself turned out to be part of the solution for a future factory, and in a way, 'generated' the solution. For example, 'if the participants find the process meaningful, they are more likely to adapt and agree upon the outcome of the process' (Wikberg-Nilsson and Johansson 2010, pp. 7–8).

7.2.2   THE USE OF DESIGN PROCESSES IN THE MINING INDUSTRY

The use of iterative and human-centric design processes is less common in mining. Horberry et al. (2018) noted that there is a limited number of organizations and groups that work actively with human-centred design; the field is in general small. They also suggested that most mining technologies have been developed from a technology-centred perspective. The problems inherent in this perspective have been noted in previous chapters, and Horberry et al. (2018) recommended the further and widespread adaptation of human-centred and participative design processes in mining. Still, there are examples of the successful application of this type of approach. Many of these come from publications by Horberry and colleagues (e.g. Horberry 2012; Horberry and Burgess-Limerick 2015; Horberry et al. 2018, 2011). These publications show the applicability of this type of thinking even to the mining industry, though there are barriers to overcome (cf. Horberry et al. 2018). The contributions of these researchers are important, but their focus is mainly on mining equipment and technology, whereas we focus on mining workplaces. Though closely related, we propose that each area needs somewhat different approaches. This can be seen in Broms and Lindahl (2005), for example. They used their approach to design LKAB's new control centre (later, they were also involved in designing control rooms for Boliden); while a human-centred approach, it is different from the approach presented by Horberry et al. (2018). On this ground, we detail below what we claim is required of a planning and design process for the design of ergonomic, safe, and attractive mining workplaces.

7.3 THE ITERATIVE PLANNING AND DESIGN PROCESS

Ranhagen (1995) presented an iterative design process that has seen successful application with good results in large projects such as for base-industry and city planning. Here, we adapt and describe a process that is adapted from Ranhagen (1995). The iterative approach consists of four steps: (1) planning the project; (2) making a diagnosis of the present status; (3) defining requirements; and (4) creating and evaluating proposals. (Ranhagen 1995 listed six steps; we have simplified them into four.) Each step is repeated – iterated – several times (see Figure 7.1). First, focus is on outline planning. Then, the focus is moved step-by-step to a more detailed plan. By systematically moving back and forth several times, the final proposals are gradually improved. We argue that this is different from many other development processes. And this constitutes the basis of what we refer to as the iterative planning and design process. While other development processes tend to be iterative, they mainly acknowledge that certain project stages have to be revisited. In the approach presented here, each project step is *systematically* revisited.

The following is a short example of how this can look in practice. Steps two to four are performed three times: first, to create preliminary proposals; then, to refine them; and lastly, to arrive at a final proposal. The first developed proposals result in new input for the present status and requirements. The focus is shifted depending on how difficult and uncertain the planning conditions are; complex situations or projects require iterations that are more detailed. In another example, a feasibility study is 'completed' three times with a focus that is gradually moved forward. This keeps the design work on track and gradually develops both detailed specifications of demands as well as detailed suggestions for well-functioning solutions from a holistic perspective. The specification of demands (see below) gets more precise for every iteration. This is especially useful for the work with subsequent, more detailed designs.

An important aspect of the iterative planning and design process is to create and judge several alternative solutions. By creating many conceptual solutions and eliminating those that do not fulfil the requirements, the number of solutions become successively fewer until a suitable concept is found. This concept is then designed in detail. In this way, the most important parts are designed and the biggest problems

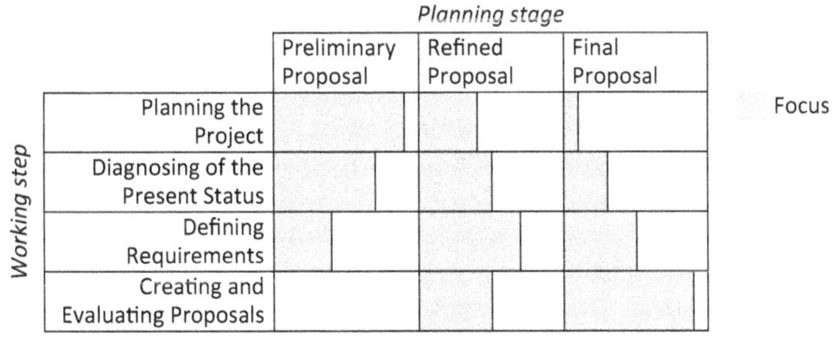

FIGURE 7.1 An example of the iterative nature of the design process. (Based on Ranhagen 1995; Johansson and Johansson 2014.)

are solved early. Working out the details is usually very time-consuming, so by focusing on the most important parts first and then on the details, solutions and designs that are less expensive and more flexible, compared to using a traditional approach, can be designed. The process puts focus on the early project stages in which fewer resources are required and it is easy to change things. As the project progresses, this changes: more resources are needed and making changes can be difficult (see Figure 1.1). In the implementation stage, it is very difficult to change things.

The process is applicable to a wide variety of projects, including the design of the mine itself, even if our focus is on workplaces. In favour of this wider applicability, we do not focus on details – these will have to be adapted to the specific circumstance – but rather on the larger principles.

7.3.1 PLANNING THE PROJECT

The planning of the project is the first step in the iterative planning and design process. It has three stages: (1) defining the project objectives; (2) forming the project organization, and (3) splitting the work into stages.

7.3.1.1 Defining the Project Objectives

A project should start by defining its objectives. When defining these objectives, they should be expressed in general terms so that they do not have to be significantly changed during the project (though some changes and revisions are always needed). A project can have more than one objective. Some examples are:

- *Production and productivity.* To increase production capacity by x tons per year over a y year period.
- *Product quality.* To achieve quality improvements in the form of higher percentage of recovered ore for every mined ton.
- *Health and safety.* To halve the amount of workplace accidents and decrease the days of sick leave from x to y.
- *Work environment.* To decrease exposure to silica and diesel fumes. To create flexible, silent, clean, and safe workplaces.
- *Environmental issues.* To lessen ground subsidence in area x.
- *Economics.* To decrease costs through rationalizations in the hoist.

We are especially concerned with the health, safety, and work environment goals. While these goals often go hand in hand with production, productivity, quality, and economic goals, setting these needs involvement of top management. If the top management sets clear and challenging goals, it will likely mean a commitment to them. If they do not, the opposite can happen. Additionally, health, safety, work environment-related goals should be set for all projects regardless of whether the primary focus is on these issues.

7.3.1.2 Forming a Project Organization

Each project also needs a project organization but depends on the size of the project. In a simple project, it is enough to have, for example, a decision-making body or person, a health and safety reference group, and a project group. The project group should have an experienced designer or planner and people from relevant functions. By relevant functions, we refer to the part of the mine, for example, that the project affects; this could be a certain department such as maintenance or geology.

Larger and more complex design projects require a larger and specialized project organization; it should then involve personnel from both inside and outside the company. Here, it is possible to use 'future workshops' as was done in the *Future Factory* (cf. Wikberg-Nilsson et al. 2011). A project leader is needed to ensure proposals are developed and submitted to the appropriate forums for decisions. Regardless of the type or size of the project, subgroups or special project groups for health and safety must be formed to analyse and eliminate problems as early as possible in the design process.

7.3.1.3   Splitting the Work into Stages

Many design processes tend to pre-define the stages required to get from the start of the project to a satisfactory result. Even when it is acknowledged that this is only possible to a certain extent, some main stages are usually delineated. Arguably, this is possible in smaller and specific projects, but as they grow larger and more diverse, this ability is diminished. The last stage of planning the project is about defining its stages. (Note the difference between stages and steps here; see Figure 7.1.) In particular, complex projects need to be divided into stages that gradually develop designs from preliminary concepts to final detailed solutions. The stages are the same even as the steps repeat, but with a changing focus. (Of course, as this is an iterative process, this step is also revised. This can result in the stages being modified.) For example, in the beginning, the first steps can be more important, and in the end of the project, the last steps can take priority.

Here, a question of the project organization also becomes relevant. It may not always be feasible to maintain a large project group that covers the needs of all stages throughout the project. To this end, Ranhagen (2004) suggested an organization based on networks that are able to form well-constructed work groups as demanded by the needs of the different project stages.

7.3.2   Diagnosing the Present Status

The second step is diagnosing the present status. This entails analysing the present status to establish and map current problems, solutions, and so on. It is a crucial step: it is the base for all health and safety improvements; to know what areas to improve, and how to improve them, requires good knowledge about the present status and situation. In short, it is about getting a good picture of the positives and negatives of the area that the project is concerned with. A recommended way of starting a diagnosis is to consider what problems the project is supposed to solve and then continue from there.

Some caution is required when making a diagnosis. It is important to remember that everyone involved in making the diagnosis have different backgrounds, experiences, and perceptions. This can affect the results of the diagnosis. For example, when looking at risks, it is important to be aware that while one person can judge something as dangerous, another can judge it as safe. To help offset these potential biases, there are tools available. The choice of tool depends on the purpose of the description, evaluation, or design. How the tool is used is also critical and, thus, so is the users' competence (the tools and competence of the user must match). Note also that there is no 'best' tool: no tool is perfect, and usually more than one tool needs to be used. Neumann (2006) made a comprehensive inventory of tools for description, evaluation, and design of work environment/ergonomics. The most common tool is the checklist-type tool (often computerized). Tools for description, evaluation, and design of work environment/ergonomics were found in all categories except for strategic decision-making. Neumann (2006) suggested that it might be suitable to use simulation approaches that are based on fundamental design specifications early in a design process (here, virtual reality can be an option). If a workplace is already in production, and a specific problem or question is to be addressed, then simpler tools may be more cost effective than actual simulations.

We also recommend that, when diagnosing the present status, effort is also spent looking towards the future state – similar to what was done in the *Future Factory* (cf. Wikberg-Nilsson et al. 2011). That is, to begin considering and discussing a vision for the future. This is also in line with the recommendation by Ranhagen (2004) to plan and develop with a future plan in mind. Even if those visions are beyond the project itself, they are important for ensuring any one solution does not hinder visions of the future state. However, it is important to remember that the future state, at least at this stage, should be visionary; they should not be solutions or concepts in their own right, as this is the purpose of the later stages. Thinking in terms of the future state is also helpful for defining requirements.

7.3.3   DEFINING REQUIREMENTS

The third step is defining requirements. The goal is to create *functional* and *measureable* objectives (i.e. they are not the project objectives) that make it possible to create and evaluate different solutions. These demands are called functional demands. The most important functional demands must be identified. The demands in general should be inspiring and reasonable. While at the start of a development process, demands can be relatively few and more visionary, they need to become more specific and detailed as the project progresses; if there are no demands or if existing demands are too vague, it will be difficult to produce any worthwhile results.

When formulating demands, they have to be stated in a way that opens up for several different solutions and proposals. Functional demands should say *what* and how *much* should be achieved. A functional demand should not specify *how* a certain requirement should be fulfilled; the demand should only state that it should be (or has to be) fulfilled. For example, a functional demand for air quality regarding purity (for example, concentration of carbon monoxide) can be achieved in several ways, such as by reducing emissions, dispersal, or exposure. That is, the demand should be for air quality, not necessarily for reduced emissions or isolated cabs. Additionally, laws and provisions (at least implicitly) specify demands. Often, these demands only specify minimum acceptable levels. When planning for projects, these levels must be much lower. For example, if the hygienic threshold value for 8 h of exposure to airborne dust is 5 mg/m^3, then the planned level should be less than 0.5 mg/m^3; the intention should always endeavour to exceed demands set by laws and provisions.

The demands have two purposes. The first is to guide and direct the creative work to create concepts and solutions. A few but highly important demands dealing with core problems usually guides this work. Too many demands should be avoided as it can strain the creative processes. The second purpose is to allow for the systematic evaluation of concepts, ideas, and detailed suggestions. The full list of demands is used so that a holistic, broad, and detailed evaluation can be performed. When possible, detailed demands have to be expressed in qualitative and quantitative terms. For example, if the demand is for 'very good air quality' (a qualitative requirement), this can be expressed with a quantitative demand on air purity, temperature humidity, and so on.

The demands will have importance in different ways; for example, some demands are perhaps not really demands but rather wishes. This is reflected in the *weight* of the demand. Essentially, though there are several variations of the procedure, each demand is assigned a number according to its (relative) importance. Which scale to use will depend on the particular project. It is also occasionally useful to use absolute demands. These are demands that have to be fulfilled in order to qualify as a solution in the first place. Such a demand could be that each solution must ensure protection from rock fall. If a solution does not, it will not be considered at all.

Finally, demands and weighs should not be arbitrary; they must be based on the present status and visions of the future state. It is thus a good idea to motivate each demand and its weight with reference to results of the previous step.

7.3.4 Creating and Evaluating Proposals

The fourth step is creating and evaluating proposals. This is a step that involves both innovative and analytical work. Planning and designing proposals are based on a mix of practice, science, and innovation. 'Common sense', analytical skills, and creativity are all used to create design alternatives that can be evaluated against the specified demands (that is, the demands defined in the previous step). As mentioned, the demands also guide the actual development of the concepts. At this step, there are also several useful tools that can be applied, such as different creative methods – like brainstorming, brainwriting, dark horse, and SCAMPER (Wikberg-Nilsson 2015) – or methods for systematic layout planning (Muther and Wheeler 1994).

After several proposals have been developed, they need to be evaluated. Here, an evaluation matrix is used. In the evaluation matrix, each row represents a demand. The rows are in turn cross-sectioned by the different designs and proposals. For each demand, every design or proposal is assessed with regards to how well it fulfils the demand. A score for that demand is obtained by multiplying the fulfilment number with the weight number. By summing every score for one design, a total score for that design is obtained. This is then done for every design (see Figure 7.2).

It is necessary to use an absolute scale for criteria fulfilment so that it is possible to judge if a proposal is good enough. This usually means that a proposal has to reach a certain score to be considered good enough. If a relative scale for criteria fulfilment is used, it is only possible to see which one of the proposals is the best;

Evaluation criteria, functional demans	Criteria weight	Design 1		Design 2		Design 3	
		Fulfilment	Score	Fulfilment	Score	Fulfilment	Score
High safety	4	3	12	3	12	4	16
High product quality	4	4	16	4	16	3	12
High productivity	3	2	6	4	12	2	6
Low environmental impact	3	1	3	3	9	4	12
Short implementation time	2	4	8	3	6	2	4
Short pay-back time	2	4	8	2	4	1	2
Total score:			53		59		52

Criteria weight scale	Criteria fulfilment scale
4 = Very important	4 = Completely fulfilled
3 = Important	3 = Almost fulfilled
2 = Not so important	2 = Acceptably fulfilled
1 = Unimportant	1 = Partly fulfilled
	0 = Nothing fulfilled

FIGURE 7.2   An example of an evaluation of three designs using an evaluation matrix.

it is not possible to see if a proposal is good enough. At this step, it is also prudent to evaluate the *current* situation using the evaluation matrix. On the one hand, this gives an indication of how much of an improvement each proposal engenders; on the other hand, it gives an indication as to what level the other proposals should exceed (the bare minimum should be to improve the current situation).

The scores awarded in the assessment of each proposal cannot be arbitrary; each score must be sufficiently motivated. This means that the criteria fulfilment scale has to have clear definitions. It must be obvious what has to be fulfilled to reach a certain level; there cannot be any ambiguity as to the difference between 'almost fulfilled' and 'completely fulfilled'. Sometimes, it is necessary that each demand, at least in the later project stages, has its own criteria fulfilment scale. These scales can be qualitative or quantitative. With air quality, for example, it is possible to use a quantitative scale where each level represents a certain amount of particles in the air (these numbers can be acquired through simulation). For demands that, for example, deal with work satisfaction or motivation, a more qualitative judgement may be required.

But an evaluation should never be reduced to a simple arithmetic task; the solutions with the highest points should not be automatically chosen just because it got the highest score. Planners and designers must always be sceptical. Before making a final decision, the results need to be assessed to see if they make sense, and if not, why not. If there are two alternatives with similar scores, then the most important criteria have to be investigated. The similar alternatives have to fulfil these criteria and alternatives that fail this check are excluded.

The result of an evaluation is dependent on the person who performs it. In order to get a fairer evaluation, different people with different expertise, experiences, and interests should perform the evaluation. It is preferable that some of the evaluators are not (directly) involved in the project. (Here, a network-based project organization is advantageous.) It is also possible to validate the results of an evaluation and reduce potential biases by having the evaluating group be separate from the project, though this group must involve relevant stakeholders. This makes it easier to make an informed decision and to formulate the best recommendations for the following design phases.

Ultimately, three types of decisions or recommendations can be made for a proposal or design: (a) the proposal is acceptable and can be further detailed; (b) the proposal needs improvements in a certain area or regarding a certain demand before it can be designed in detail, and (c) the proposal should be rejected and is not suitable for further detailed design. By using this systematic evaluation of design proposals, more rational, conscious, and wise decisions can be made so that the best solutions are promoted.

At this point, the 'cycle' restarts: the project goes through the working steps again to evaluate if the goals or project organization need to be updated (perhaps a new stakeholder has been identified or a new goal uncovered). Then, the present situation is again diagnosed, for example, to obtain new relevant data, the requirements are modified; new proposals are developed, modified, or redesigned, until finally, an optimal solution is found.

7.4   SUMMARY

The use of the iterative planning and design process as conceptualized here is advantageous because it is based on experiences on city planning and other large projects (cf. Ranhagen 2004, 1995). This is suitable for mining, on the one hand, because mining projects are large and complex; they need to be approached systematically but also in a way that efficiently deals with variation and complexities. On the other hand, like city planning projects, mining projects, compared to other industrial projects, usually have a significant impact on the surrounding society. And even when they do not have such an effect, when mines are located remotely and utilize a FIFO organization, the endeavour to achieve attractive workplaces still mandates an approach that manages to capture the needs, wants, and wishes of a wider group of stakeholders. The most obvious of these stakeholders, in this case, are young people , women, and so on who do not – and do not want to – work in the mining industry. By using a project approach that readily seeks to involve these stakeholders, for example, by using participative 'future workshops' (Wikberg-Nilsson et al. 2011), we can hope to better design workplaces in which they will want to work.

References

Abrahamsson, L. 2002. Restoring the order: Gender segregation as an obstacle to organizational development. *Applied Ergonomics* 33(6):549–557. doi:10.1016/S0003-6870(02)00043-1.

Abrahamsson, L. 2006. Exploring construction of gendered identities at work. In *Work, Subjectivity and Learning: Understanding Learning Through Working Life*, eds. S. Billet, T. Fenwick, and M. Somerville, 105–121. Dordrecht: Springer.

Abrahamsson, L. 2009. *Att återställa ordningen (Restoring the Order)*. Umeå: Borea.

Abrahamsson, L. 2014. Gender and the modern organization: Ten years after. *Nordic Journal of Working Life Studies* 4(4):109–136.

Abrahamsson, L. and J. Johansson. 2006. From grounded skills to sky qualifications: a study of workers creating and recreating qualifications, identity and gender at an underground Iron Ore Mine in Sweden. *Journal of Industrial Relations* 48(5):657–76. doi:10.1177/0022185606070110.

Abrahamsson, L. and M. Somerville. 2007. Changing storylines and masculine bodies in Australian coal mining organisations. *Nordic Journal for Masculinity Studies* 2(1):52–69.

Abrahamsson, L., B. Johansson, and J. Johansson. 2009. Future of metal mining: Sixteen predictions. *International Journal of Mining and Mineral Engineering* 1(3):304–312. doi:10.1504/IJMME.2009.027259.

Abrahamsson, L., E. Segerstedt, M. Nygren, J. Johansson, B. Johansson, I. Edman and A. Åkerlund. 2014. *Mining and Sustainable Development: Gender, Diversity and Work Conditions in Mining*. Luleå: Luleå University of Technology.

Abrahamsson, L., J. Lööw, M. Nygren and E. Segerstedt. 2015. How to get at social licence to mine. Paper presented at *Third International Future Mining Conference*, 2015, 4–6 November, Sydney.

Abrahamsson, L., J. Lööw, M. Nygren and E. Segerstedt. 2016. Challenges in obtaining a social licence to mine. *AusIMM Bulletin* 6. www.ausimmbulletin.com/feature/challenges-in-obtaining-a-social-licence-to-mine/.

Adler, P. 1995. Democratic taylorism: The Toyota production system at NUMMI. In *Lean Work: Empowerment and Exploitation in the Global Auto Industry*, ed. S. Babson, 207–219. Detroit, MI: Wayne State University Press.

Ail, K. K. and E. Y. Baffi. 2007. Environmental impact of artisanal and small scale gold mining in developing countries. Paper presented at *The Sixteenth International Symposium on Mine Planning and Equipment Selection*, 11–13 December, Bangkok.

Alam, M. M. and E. B. Hamida. 2014. Surveying wearable human assistive technology for life and safety critical applications. *Sensors* 14(5):9153–9209. doi:10.3390/s140509153.

Albanese, T. and J. McGagh. 2011. Future trends in mining. In *SME Mining Engineering Handbook*, ed. P. Darling, vol. 1, 3rd ed.: 21–36. Englewood, CO: Society for Mining, Metallurgy, and Exploration.

Andersson, E. 2012. Malmens manliga mysterium: En interaktiv studie om kön och tradition i modernt gruvarbete (The male mystery of the ore: An interactive study on gender and tradition in modern mining work). *Ph.D. Dissertation*, Luleå University of Technology.

Angelis, J., R. Conti, C. Cooper and C. Gill. 2011. Building a high-commitment lean culture. *Journal of Manufacturing Technology Management* 22(5):569–586. doi:10.1108/17410381111134446.

Åteg, M. 2006. Aktiviteter och lärande för attraktivt arbete: utvecklingsprocesser inom verk-
stadsindustrin (Activities and learning for attractive work: development processes in
the engineering industry). *Ph.D. Dissertation*, Royal Institute of Technology/Dalarna
University.

Åteg, M., A. Hedlund and B. Pontén. 2004. *Attraktivt arbete: Från anställdas uttalanden till
skapandet av en modell (Attractive Work: From Employees' Statements to the Creation
of a Model)*. Arbetsliv i omvandling 2004:1. Stockholm: Arbetslivsinstitutet.

Åteg, M., A. Hedlund, and B. Pontén. 2004. Attraktivt arbete: Från anställdas uttalanden till
skapandet av en modell (Attractive work: From employees' statements to the creation
of a model). *Arbetsliv i omvandling* 1:1–67.

Atlas Copco. 2014. *Underground Mining: A Global Review of Methods and Practices*.
Örebro: Atlas Copco Rock Drills.

Bäckblom, G. 2009. Mine of the Future (MIFU) – Account from the student event LTU, Nov
16th 2009. Report/memo of WP2 of SMIFU. Luleå: Rock Tech Centre.

Bäckblom, G., E. Forssberg, S. Haugen, J. Johansson, T. Naarttijärvi, and B. Öhlander. 2010.
Smart Mine of the Future Conceptual Study 2009–2010: Final report. Luleå: Rock
Tech Centre.

Bahn, S. 2013. Workplace hazard identification and management: The case of an underground
mining operation. *Safety Science* 57:129–137. doi:10.1016/j.ssci.2013.01.010.

Bainbridge, L. 1983. Ironies of automation. *Automatica* 19(6):775–79.
doi:10.1016/0005-1098(83)90046-8.

Banks, J. A. 1965. Alienation and freedom, book review. *The British Journal of Sociology*
16(1):90–91.

Bartos, P. J. 2007. Is mining a high-tech industry? Investigations into innovation and productivity advance. *Resources Policy* 32(4):149–158. doi:10.1016/j.resourpol.2007.07.001.

Bassan, J., V. Srinivasan, P. Knights and C. Farrelly. 2008. A day in the life of a mine
worker 2025. Paper presented at *First International Future Mining Conference*, 19–21
November, Sydney.

Bellamy, D. and L. Pravica. 2011. Assessing the impact of driverless haul trucks in Australian
surface mining. *Resources Policy* 36:149–158. doi:10.1016/j.resourpol.2010.09.002.

Berggren, C., T. Björkman and E. Hollander. 1991. *Are They Unbeatable? Report from a
Field Trip to Study Transplants, the Japanese Owned Auto Plants in North America*.
Stockholm: Royal Institute of Technology.

Berglund, R. 2006. *Smart Lean – Möjligheter att utnyttja Lean-konceptet för att skapa
en god arbetssituation (Smart Lean – Opportunities for Using the Lean-Concept
for Creating Good Working Conditions)*. Lund: Institutet för verkstadsteknisk
forskning.

Berner, B. 2003. *Vem tillhör tekniken? Kunskap och kön i teknikens värld (To Whom Does
Technology Belong? Knowledge and Gender in the World of Technology)*. Lund:
Arkiv.

Beus, J. M., M. A. McCord and D. Zohar. 2016. Workplace safety: A review and research synthesis. *Organizational Psychology Review* 6(4):352–81. doi:10.1177/2041386615626243.

Bhamu, J. and K. Singh Sangwan. 2014. Lean manufacturing: Literature review and research
issues. *International Journal of Operations & Production Management* 34(7):876–
940. doi:10.1108/IJOPM-08-2012-0315.

Blank, V. L. G., F. Diderichsen and R. Andersson. 1996. Technological development
and occupational accidents as a conditional relationship: A study of over eighty
years in the Swedish mining industry. *Journal of Safety Research* 27(3):137–46.
doi:10.1016/0022-4375(96)00014-X.

Blank, V. L. G., L. Laflamme, F. Diderichsen and R. Andersson. 1998. Choice of a denomi-
nator for occupational injury rates: A study of the development of a Swedish iron-ore
mine. *Journal of Safety Research* 29(4):263–73. doi:10.1016/s0022-4375(98)00052-8.

Blank, V. L. G., R. Andersson, A. Lindén and B.-C. Nilsson. 1995. Hidden accident rates and patterns in the Swedish mining industry due to involvement of contractor workers. *Safety Science* 21(1):23–35. doi:10.1016/0925-7535(95)00004-6. University of Chicago Press.

Blumstein, D., R. Ferriter, J. Powers, and M. Reiher. 2011. *Accidents – The Total Cost: A Guide for Estimating the Total Cost of Accidents.* Golden, CO: Colorado School of Mines.

Bohgard, M., R. Akselsson, I. Holmér, G. Johansson, F. Rassner and L. Swensson. 2008. Physical factors. In *Work and Technology on Human Terms*, ed. M. Bohgard, S. Karlsson, E. Lovén, L. Mikaelsson, L. Mårtensson, A. Osvalder, et al., 191–307. Stockholm: Prevent.

Bohgard, M., S. Karlsson, E. Lovén, L.-Å. Mikaelsson, L. Mårtensson, A.-L. Osvalder, L. Rose and P. Ulfvengren (eds.). 2009. *Work and Technology on Human Terms.* Stockholm: Prevent.

Bolis, I., C. M. Brunoro and L. I. Sznelwar. 2014. Mapping the relationships between work and sustainability and the opportunities for ergonomic action. *Applied Ergonomics* 45(4): 1225–1239. doi:10.1016/j.apergo.2014.02.011.

Brand, S. 1990. Arbetsskador i svenska gruvor (Occupational injuries in Swedish mines). In *Från yrkesfara till arbetsmiljö: Yrkesinspektionen 100 år (From occupational hazard to work environment: 100 years of the labor inspectorate)*, ed. Ö. Ekström, and I. Hall, 172–178. Solna: Arbetarskyddsstyrelsen.

Broms, G. and P. Lindahl. 2005. *Effektivt arbete i processindustrin: Hur man gör. Från strategi till genomförande (Effective Work in the Process Industry: How to Do It. From Strategy to Realization).* Stockholm: Vinnova.

Burgess-Limerick, R., J. Joy, T. Cooke and T. Horberry. 2012. EDEEP – An innovative process for improving the safety of mining equipment. *Minerals* 2:272–282. doi:10.3390/min2040272

Cameron, I. and B. Hare. 2008. Planning tools for integrating health and safety in construction. *Construction Management and Economics* 26(9):899–909. doi:10.1080/01446190802175660.

Carayon, P. and M. J. Smith. 2000. Work organization and ergonomics. *Applied Ergonomics* 31 (6):649–62. doi:10.1016/S0003-6870(00)00040-5.

Castillo, G., L. Alarcón and V. González. 2015. Implementing lean production in copper mining development projects: Case study. *Journal of Construction Engineering and Management* 141(1). doi:10.1061/(ASCE)CO.1943-7862.0000917.

Cavender, B. 2000. Rethinking the use of new technology to improve operational performance. *Mining Engineering* 52(12):61–67.

Claassen, J. 2016. Application of manufacturing management and improvement methodologies in the Southern African mining industry. *Journal of the Southern African Institute of Mining and Metallurgy* 116(2):139–148. doi:10.17159/2411-9717/2016/v116n2a5.

Cliff, D. and T. Horberry. 2008. Hours of work risk factors for coal mining. *International Journal of Mining and Mineral Engineering* 1(1):77–94. doi:10.1136/oem.2004.016667.

Connell, R. W. 1995. *Masculinities.* Los Angeles, CA: University of California Press.

Cox, L. A. 2008. What's wrong with risk matrices? *Risk Analysis* 28(2):497–512. doi:10.1111/j.1539-6924.2008.01030.x

Dahlström, E., B. Gardell, B. G. Rundblad, B. Wingårdh and J. Hallin. 1966. *Teknisk förändring och arbetsanpassning (Technological change and work adaptation).* Stockholm: Prisma.

Dembe, A. E., J. B. Erickson, R. G. Delbos and S. M. Banks. 2005. The impact of overtime and long work hours on occupational injuries and illnesses: New evidence from the United States. *Occupational and Environmental Medicine* 62(9):588–597. doi:10.1136/oem.2004.016667.

Demiral, Y. and A. Ertürk. 2013. Safety and health in mining in Turkey. In *Occupational Safety and Health in Mining: Anthology on the Situation in 16 Mining Countries*, ed. K. Elgstrand, and E. Vingård, 87–93. Gothenbutg: University of Gothenburg. http://hdl.handle.net/2077/32882. Arbete och Hälsa 47(2).

Dencker, K., J. Stahre, P. Gröndahl, L. Mårtensson, T. Lundholm and C. Johansson. 2007. An approach to proactive assembly systems: Towards competitive assembly systems. Paper presented at *International Symposium on Assembly and Manufacturing*, 22–25 July, Michigan.

Dombrowski, U. and T. Wagner. 2014. Mental strain as field of action in the 4th industrial revolution. *Procedia CIRP* 17:100–105. doi:10.1016/j.procir.2014.01.077.

Donoghue, A. M. 2004. Occupational health hazards in mining: An overview. *Occupational Medicine* 54(5):283–89. doi:10.1093/occmed/kqh072.

Dunstan, K., B. Lavin and R. Sanford. 2006. The application of lean manufacturing in a mining environment. In: *International Mine Management Conference*, Melbourne, 16–17 October 2006, The Australasian Institute of Mining and Metallurgy, pp. 145–154.

Ekevall, E., B. Gillespi and L. Riege. 2008. *Improving Safety Performance in the Australian Mining Industry Through Enhanced Reporting*. Brisbane: PricewaterhouseCoopers.

Elgstrand, K. and E. Vingård, eds. 2013. *Occupational Safety and Health in Mining Anthology on the Situation in 16 Mining Countries*. Gothenburg: University of Gothenburg. http://hdl.handle.net/2077/32882. *Arbete och Hälsa* 47(2).

Elgstrand, K. and E. Vingård. 2013. Safety and health in mining. In *Occupational Safety and Health in Mining*: *Anthology on the Situation in 16 Mining countries*, eds. K. Elgstrand, and E. Vingård: 1–14. Gothenburg: University of Gothenburg. http://hdl.handle.net/2077/32882. Arbete och Hälsa 47(2).

Ellström, P.-E. 2006. The meaning and role of reflection in informal learning at work. In *Productive Reflection at Work: Learning for Changing Organizations*, eds. D. Boud, P. Cressey and P. Docherty. London: Routledge.

Ely, R. J. and D. Meyerson. 2008. Unmasking manly men. *Harvard Business Review* 86 (7/8):1–3.

Eriksson, U. 1991. Gruva och arbete: Kiirunavaara 1890–1990 (Mining and miners: Kiirunavaara 1890–1990), vol. 3, 1950–1970. *Ph.D. Dissertation*, Uppsala University.

European Commission. 2010. *Health and Safety at Work in Europe (1999–2007): A Statistical Portrait*. Luxembourg: Publications Office of the European Union.

European Commission. 2011. *Socio-Economic Costs of Accidents at Work and Work-Related Ill Health: Key Messages and Case Studies*. Luxembourg: Publications Office of the European Union.

Eveline, J. and M. Booth. 2002. Gender and sexuality in discourses of managerial control: The case of women miners. *Gender, Work and Organization* 9(5):556–578. doi:10.1111/1468-0432.00175.

Feickert, E. 2013. Safety and health in mining in China. In *Occupational Safety and Health in Mining: Anthology on the Situation in 16 Mining Countries*, ed. K. Elgstrand, and E. Vingård, 23–30. Gothenbutg: University of Gothenburg. http://hdl.handle.net/2077/32882. Arbete och Hälsa 47(2).

Feki, M. A., F. Kawsar, M. Boussard, and L. Trappeniers. 2013. The Internet of Things: The next technological revolution. *Computer* 46(2):24–25. doi:10.1109/MC.2013.63.

Fiscor, S. 2014. Three steps to improving mine performance. *Engineering & Mining Journal* 215(6):92–101.

Franzen, C., P. Lärkeryd, S. Sjölander and J. Borgström. 2010. *Det lönar sig: genusmedveten ledning och styrning i verkstadsindustri (Gender Conscious Management and Control in the Manufacturing Industry)*. Uppsala: Näringslivets Ledarskapsakademi.

Frick, K. 2000. *Systematic Occupational Health and Safety Management: Perspectives on an International Development*. Amsterdam: Pergamon.

Frick, K. and J. Wren. 2000. Reviewing occupational health and safety management—Multiple roots, diverse perspectives and ambiguous outcomes. In *Systematic Occupational Health and Safety Management Perspectives on an International Development*, ed. K. Frick, 17–42. Amsterdam: Pergamon.

Frick, K., P. L. Jensen, M. Quinlan and T. Wilthagen. 2000. Systematic occupational health and safety management – An introduction to a new strategy for occupational safety, health and well-being. In *Systematic Occupational Health and Safety Management Perspectives on an International Development*, ed. K. Frick, 1–14. Amsterdam: Pergamon.

Galdón-Sánchez, J. E. and J. A. Schmitz Jr. 2002. Competitive pressure and labor productivity: world Iron-Ore Markets in the 1980's. *The American Economic Review* 92(4):1222–1235. www.jstor.org/stable/3083310.

Gambatese, J. A. 2000. Safety in a designer's hands. *Civil Engineering* 70(6):56–59.

Gambatese, J. A., M. Behm and S. Rajendran. 2008. Design's role in construction accident causality and prevention: Perspectives from an expert panel. *Safety Science* 46(4):675–691. doi:10.1016/j.ssci.2007.06.010.

Gardell, B. 1976. *Arbetsinnehåll och livskvalitet: En sammanställning och diskussion av samhällsvetenskaplig forskning rörande människan och arbetet (Work Content and Quality of Life: A Compilation and Discussion of Social Science Research on Humans and Work).* Stockholm: Prisma.

Geological Survey of Sweden. 2016. Statistics of the Swedish Mining Industry 2015. Periodiska publikationer 2016, p. 1. Uppsala: Geological Survey of Sweden.

Gherardi, S. and D. Nicolini. 2000. The organizational learning of safety in communities of practice. *Journal of Management Inquiry* 9(1):7–18.

Gill, S. 2014. Industry prepares for the next industrial revolution. *Control Engineering* 27th of June 2013. www.controleng.com/single-article/industry-prepares-for-the-next-industrial-revolution.

Goodman, P. S. and S. Garber. 1988. Absenteeism and accidents in a dangerous environment: Empirical analysis of underground coal mines. *Journal of Applied Psychology* 73(1):81–86. doi:10.1037/0021-9010.73.1.81.

Groves, W. A., V. J. Kecojevic and D. Komljenovic. 2007. Analysis of fatalities and injuries involving mining equipment. *Journal of Safety Research* 38(4):461–70. doi:10.1016/j.jsr.2007.03.011.

Haddon, W., Jr. 1963. A note concerning accident theory and research with special reference to motor vehicle accidents. *Annals of the New York Academy of Sciences* 107(2):635–46. doi:10.1111/j.1749-6632.1963.tb13307.x.

Haddon, W., Jr. 1973. Energy damage and the 10 countermeasure strategies. *Journal of Trauma* 13:321–331.

Haddon, W., Jr. 1995. Energy damage and the 10 countermeasure strategies. *Injury Prevention* 1:40–44.

Haldane, S. 2013. Safety and health in mining in Canada. In *Occupational Safety and Health in Mining: Anthology on the Situation in 16 Mining Countries*, ed. K. Elgstrand, and E. Vingård, 129–136. Gothenbutg: University of Gothenburg. http://hdl.handle.net/2077/32882. Arbete och Hälsa 47(2).

Harms-Ringdahl, L. 2004. Riskhantering i företag och samhälle (Risk management in companies and society). In *Framtidens arbetsmiljö- och tillsynsarbete (Work Environmental and Supervisory Work of the Future)*, ed. B. Johansson, K. Frick, and J. Johansson, 310–327. Lund: Studentlitteratur.

Harms-Ringdahl, L. 2013. *Guide to Safety Analysis for Accident Prevention*. Stockholm: IRS Riskhantering. www.irisk.se/sabook/SA-book1.pdf.

Hartman, H. L. and J. M. Mutmansky. 2002. *Introductory Mining Engineering*. 2nd ed. Hoboken, NJ: John Wiley and Sons.

Haslam, R. and P. Waterson. 2013. Ergonomics and sustainability. *Ergonomics* 56(3):343–347. doi:10.1080/00140139.2013.786555.

Hattingh, T. S. and O. T. Keys. 2010. How applicable is industrial engineering in mining. In: *4th International Platinum Conference: Platinum in Transition 'Boom or Bust'*, Sun City, South Africa, 11–14 October 2010, The Southern African Institute of Mining and Metallurgy, Johannesburg, pp. 205–210.

Haugen, S. 2013. Lean mining. *Mineralproduksjon* 3:B21–B40.

Hebblewhite, B. 2008. Education and training for the international mining industry – Future challenges and opportunities. Paper presented at: *First International Future Mining Conference*, 19–21 November, Sydney.

Hedlund, A. 2006. The attractiveness of the work is affected when production of handcrafted log houses moves indoors. *Silva Fennica* 40(3):545–558. doi:10.14214/sf.336.

Hedlund, A. 2007. Attraktivitetens dynamik: studier av förändringar i arbetets attraktivitet (The dynamic of attractiveness: studies of changes in the attractiveness of work). *Ph.D. Dissertation*, Royal Institute of Technology/Dalarna University.

Hedlund, A. and B. Pontén. 2006. *Införande av systematiskt arbetsmiljöarbete på träföretag: Utvärdering av en metod, dess resultat och påverkan på arbetets attraktivitet (Implementation of a Systematic Work Environment Management System in Woodworking Companies: Evaluation of a Method, Its Results and the Influence on the Work Attractiveness)*. Arbete och Hälsa 15. Stockholm: National Institute for Working Life.

Hedlund, A., I.-M. Andersson and G. Rosén. 2010. Är dagens arbeten attraktiva? Värderingar hos 1 400 anställda (Are today's jobs attractive? Values of 1,440 employees). *Arbetsmarknad & Arbetsliv* 16(4):31–44. www.intra.kau.se/dokument/upload/C10B99370f047252A6txn1B5ABB9/Hedlund_tryckt.pdf.

Hedlund, A., M. Åteg, I.-M. Andersson and G. Rosén. 2010b. Assessing motivation for work environment improvements: Internal consistency, reliability and factorial structure. *Journal of Safety Research* 41(2):145–151. doi:10.1016/j.jsr.2009.12.005.

Helgeson, B. and P. Bergman. 1983. *Teknisk utveckling och industriarbetets villkor: Kern & Schumanns studie "Industriearbeit und Arbeiterbewusstein" (Technological Development and the Conditions of Industrial Work: Kern & Schumann's Study "Industriearbeit und Arbeiterbewusstein")*. Luleå: Tekniska högskolan i Luleå.

Helman, J. 2012. Analysis of the potentials of adapting elements of lean methodology to the unstable conditions in the mining industry. *AGH Journal of Mining and Geoengineering* 36:151–157.

Herzberg, F. 1968. One more time: How do you motivate employees? *Harvard Business Review* 46(1): 53–62.

Hines, P., M. Holweg and N. Rich. 2004. Learning to evolve: A review of contemporary lean thinking. *International Journal of Operations & Production Management* 24(10):994–1011. doi:10.1108/01443570410558049.

Hollnagel, E. 2007. From overcoming limitations to securing safety: Defining the new mission of human-machine systems. Paper presented at *10th IFAC/IFIP/IFORS/IEA Symposium on Analysis, Design, and Evaluation of Human-Machine Systems*, 4–6 September, Seoul.

Hood, M. 2004. Advances in hard rock mining technology. In *Proceedings of the Mineral Economics and Management Society*, 13th Annual Conference, 21–23.

Hoonakker, P. and C. Korunka. 2014. Introduction. In C. Korunka and P. Hoonakker, eds., *The Impact of ICT on Quality of Working Life*. Dordrecht: Springer.

Hopkins, A. 2004. Quantitative risk assessment: A critique. Working Paper 25. National Research Centre for OHS Regulation, Australian National University.

Hopkins, A. 2009. Thinking about process safety indicators. *Safety Science* 47(4):460–465. doi:10.1016/j.ssci.2007.12.006

Horberry T. and R. Burgess-Limerick. 2015. Applying a human-centred process to re-design equipment and work environments. *Safety* 1(1):7–15. doi:10.3390/safety1010007.

Horberry T., R. Burgess-Limerick and L. J. Steiner. 2011. *Human Factors for the Design, Operation, and Maintenance of Mining Equipment.* Boca Raton, FL: CRC Press.

Horberry, T. 2012. The health and safety benefits of new technologies in mining: A review and strategy for designing and deploying effective user-centred systems. *Minerals* 2(4):417–425. doi:10.3390/min2040417.

Horberry, T. J., R. Burgess-Limerick and L. J. Steiner. 2011. *Human Factors for the Design, Operation, and Maintenance of Mining Equipment.* Boca Raton, FL: CRC Press.

Horberry, T., R. Burgess-Limerick and L. J. Steiner. 2018. *Human-Centered Design for Mining Equipment and New Technology.* Boca Raton, FL: CRC Press.

Horberry, T., R. Burgess-Limerick and R. Fuller. 2013. The contributions of human factors and ergonomics to a sustainable minerals industry. *Ergonomics* 56(3):556–564. doi:10.1080/00140139.2012.718800.

Hunter, S. L. 2008. The toyota production system applied to the upholstery furniture manufacturing industry. *Materials and Manufacturing Processes* 23(7):629–634. doi:10.1080/10426910802316476.

Hutchings, K., H. de Cieri and T. Shea. 2011. Employee attraction and retention in the Australian resources sector. *Journal of Industrial Relations* 53(1):83–101. doi:10.1177/0022185610390299.

Imai, M. 1986. *Kaizen: The Key to Japan's Competitive Success.* New York: Random House.

International Council on Mining & Metals (ICMM). 2012. *Overview of Leading Indicators for Occupational Health and Safety in Mining.* London: ICMM.

International Council on Mining & Metals (ICMM). n.d. *Benchmarking 2016 safety data: progress of ICMM members* (webpage). Available at: www.icmm.com/safety-data-2016 (Accessed 2018-05-07).

International Ergonomics Association (IEA). 2017. Definition and domains of ergonomics. www.iea.cc/whats/ (Accessed 2017-10-25).

International Labour Organization (ILO). 2006. *Occupational Safety and Health: Synergies between Security and Productivity.* Geneva: International Labour Organization.

Jäderblom, N. 2017. From diesel to battery power in underground mines: A pilot study of diesel free LHDs. *Master's Thesis*, Luleå University of Technology.

Järvholm, B. 2013. Safety and health in mining in Sweden. In *Occupational Safety and Health in Mining: Anthology on the Situation in 16 Mining Countries*, eds. K. Elgstrand, and E. Vingård: 77–86. Gothenburg: University of Gotenburg. http://hdl.handle.net/2077/32882. Arbete och Hälsa 47(2).

Johansson, B. and J. Johansson. 2008. *Work Environment and Work Organization in the Swedish and Finnish Mining Industry.* Luleå: Department of Human Work Sciences.

Johansson, B. and J. Johansson. 2014. 'The New Attractive Mine': 36 research areas for attractive workplaces in future deep metal mining. *International Journal of Mining and Mineral Engineering* 5(4):350–361. doi:10.1504/IJMME.2014.066582.

Johansson, B., J. Johansson and L. Abrahamsson. 2010a. Attractive workplaces in the mine of the future: 26 statements. *International Journal of Mining and Mineral Engineering* 2(3):239–252. doi:10.1504/IJMME.2010.037626

Johansson, B., K. Rask and M. Stenberg. 2010b. Piece rates and their effects on health and safety – A literature review. *Applied Ergonomics* 41(4):607–614. doi:10.1016/j.apergo.2009.12.020.

Johansson, J. 1986. Teknisk och organisatorisk gestaltning: exemplet LKAB (Technological and organizational Gestaltung: The LKAB example). *Ph.D. Dissertation*, Luleå University of Technology.

Johansson, J. and L. Abrahamsson. 2009. The good work: Swedish trade union vision in the shadow of lean production. *Applied Ergonomics* 40(4):775–780. doi:10.1016/j. apergo.2008.08.001.

Johansson, J., B. Johansson, J. Lööw, M. Nygren and L. Abrahamsson. 2018. Attracting young people to the mining industry: Six recommendations. *International Journal of Mining and Mineral Engineering* 9(2):94–108. doi:10.1504/IJMME.2018.091967.

Johansson, J., L. Abrahamsson, B. Bergvall Kåreborn, Y. Fältholm, C. Grane, and A. Wykowska. 2017. Work and organization in a digital context. *Management Revue* 28(3):281–297. doi:10.5771/0935-9915-2017-3-281.

Joy, J. 2004. Occupational safety risk management in Australian mining. *Occupational Medicine* 54(5):311–315. doi:10.1093/occmed/kqh074

Jürgens, U. 1997. Rolling back cycle times: The renaissance of the classic assembly line in final assembly. In *Transforming Automobile Assembly*, eds. K. Shimokawa, U. Jürgens, and T. Fujimoto, Berlin: Springer.

Kagerman, H., W. Wahlster and J. Helbig. 2013. *Recommendations for Implementing the Strategic Initiative Industry 4.0.* München: Acatech.

Karasek, R. and T. Theorell. 1990. *Healthy Work: Stress, Productivity and the Reconstruction of Working Life.* New York, NY: Basic Books.

Katen, K. P. 1992. Health and safety standards. In *SME Mining Engineering Handbook*, ed. H. L. Hartman, 2nd ed., 162–173. Littleton, CO: Society for Mining, Metallurgy, and Exploration.

Kecojevic, V. and M. Radomsky. 2004. The causes and control of loader and truck-related fatalities in surface mining operations. *Injury Control and Safety Promotion* 11(4):239–251. doi:10.1080/156609704/233/289779.

Kecojevic, V., D. Komljenovic, W. Groves and M. Radomsky. 2007. An analysis of equipment-related fatal accidents in U.S. mining operations: 1995–2005. *Safety Science* 45(8):864–874. doi:10.1016/j.ssci.2006.08.024.

Kern, H. and M. Schumann. 1974. *Industriearbeit und Arbeiterbewusstsein (Industrial work and worker awareness).* Frankfurt am Main: Europaische Verlagsanstalt.

Kern, H. and M. Schumann. 1987. Limits of the division of labour: New production and employment concepts in West German industry. *Economic and Industrial Democracy* 8(2):151–170. doi:10.1177/0143831X8782002.

Khaba, S. and C. Bhar. 2017. Quantifying SWOT analysis for the Indian coal mining industry using fuzzy DEMATEL. *Benchmarking: An International Journal* 24(4):882–902. doi:10.1108/BIJ-06-2016–0089.

Kissel, F. N. 2003. Dust control methods in tunnels and underground mines. In *Handbook for Dust Control in Mining*, ed. F. N. Kissel, 3–13. Pittsburgh, PA: National Institute for Occupational Safety and Health. www.cdc.gov/niosh/mining/UserFiles/works/pdfs/2003-147.pdf.

Kizil, M. S. and W. R. Hancock. 2008. Internet-based remote machinery control. Paper presented at the *First International Future Mining Conference*, 19–21 November, Sydney.

Klippel, A. F., C. O. Petter and J. A. V. Antunes Jr. 2008. Lean management implementation in mining industries. *Dyna* 75(154):81–89.

Kopacek, P. 2015. Automation and TECIS. *IFAC-Papers-OnLine* 48(24):21–27. doi:10.1016/j. ifacol.2015.12.050.

Kronlund, J., J. Carlsson, I.-L. Jensen and C. Sundström-Frisk. 1973. *Demokrati utan makt: LKAB efter strejken (Democracy without power: LKAB after the strike).* Stockholm: Prisma.

Krzemień, S. and A. Krzemień. 2013. Safety and health in mining in Poland. In *Occupational Safety and Health in Mining: Anthology on the Situation in 16 mining countries*, ed. K. Elgstrand and E. Vingård, 59–66. Gothenburg: University of Gothenburg. http://hdl. handle.net/2077/32882. Arbete och Hälsa 47(2).

Kurdve, M. 2014. Development of collaborative green lean production systems. *Ph.D. Dissertation*, Mälardalen University.

Laflamme, L. and V. L. G. Blank. 1996. Age-related accident risks: Longitudinal study of Swedish iron ore miners. *American Journal of Industrial Medicine* 30(4):479–487. doi:10.1002/(SICI)1097-0274(199610)30:4<479::AID-AJIM14>3.0.CO;2-1.

Lahiri-Dutt, K. 2007. Roles and status of women in extractive industries in India: Making a place for a gender-sensitive mining development. *Social Change* 37(4):37–64. doi:10.1177/004908570703700403.

Lahiri-Dutt, K. 2012. Digging women: Towards a new agenda for feminist critiques of mining. *Gender, Place and Culture* 19(2):193–212. doi:10.1080/0966369X.2011.572433.

Landsbergis, P. A., J. Cahill and P. Schnall. 1999. The impact of lean production and related new systems of work organization on worker health. *Journal of Occupational Health Psychology* 4(2):208–130.

Lasi, H., P. D. P. Fettke, H. G. Kemper, D. I. T. Feld and D. H. M. Hoffmann. 2014. Industry 4.0. *Business & Information Systems Engineering* 6(4):239–242. doi:10.1007/s12599-014-0334-4.

Laurence, D. 2011. Mine safety. In *SME Mining Engineering Handbook*, ed. P. Darling, vol. 2, 3rd ed.: 1557–1566. Englewood, CO: Society for Mining, Metallurgy, and Exploration.

Lee, G. A. 2011. Management, employee relations, and training. In *SME Mining Engineering Handbook*, ed. P. Darling, vol. 1, 3rd ed.: 317–333.Englewood, CO: Society for Mining, Metallurgy, and Exploration.

Lee, J., H. A. Kao and S. Yang. 2014. Service innovation and smart analytics for Industry 4.0 and big data environment. *Procedia CIRP* 16:3–8. doi:10.1016/j.procir.2014.02.001.

Lenné, M. G., P. M. Salmon, C. C. Liu and M. Trotter. 2012. A systems approach to accident causation in mining: An application of the HFACS method. *Accident Analysis and Prevention* 48:111–17. doi:10.1016/j.aap.2011.05.026.

Lever, P. 2011. Automation and robotics. In *SME Mining Engineering Handbook*, ed. P. Darling, vol. 1, 3rd ed.: 805–827. Englewood, CO: Society for Mining, Metallurgy, and Exploration.

Li, X., D. J. McKee, T. Horberry and M. S. Powell. 2011. The control room operator: The forgotten element in mineral process control. *Minerals Engineering* 24(8):894–902. doi:10.1016/j.mineng.2011.04.001.

Liker, J. 2004. *The Toyota Way*. New York: McGraw-Hill.

Lilley, R., A. Samaranayaka and H. Weiss. 2013. International comparison of International Labour Organisation published occupational fatal injury rates: How does New Zealand compare internationally? Commissioned report for the Independent Taskforce on Workplace Health and Safety. Dunedin: University of Otago.

LKAB. 2017. *Annual and Sustainability Report 2016*. Luleå: LKAB.

Lööw, J. 2015. *Lean Production in Mining: An Overview of Lean Production in the Mining Industry, Produced for the I2Mine (Innovative Technologies and Concepts for the Intelligent Deep Mine of the Future) Project*. Luleå: Luleå University of Technology.

Lööw, J. and J. Johansson. 2015a. An overview of lean production and its application in mining. Paper presented at: *Aachen Fifth International Mining Symposia: Mineral Resources and Mine Development*, 27–28 May, Aachen.

Lööw, J. and J. Johansson. 2015b. Work organisation for attractive mining – Lean mining and the working environment. Paper presented at: *Third International Future Mining Conference*, 4–6 November, Sydney.

Lööw, J., C. Grane, N. Jäderblom, J. Lundgren and J. Johansson. 2017. A baseline study for each technology project. *Report from WP8 of the SIMS project*. Luleå.

Lööw, J., M. Nygren and J. Johansson. 2017. *Säkerhet i svensk gruvindustri: 30 år av sänkta olycksfallsfrekvenser-och den fortsatta vägen framåt (Safety in the Swedish mining industry: 30 years of reduced accident frequency rates-and the way forward)*. Luleå: Luleå University of Technology. http://urn.kb.se/resolve?urn=urn:nbn:se:ltu:diva-63440.

Losonci, D., K. Demeter and I. Jenei. 2011. Factors influencing employee perceptions in lean transformations. *International Journal of Production Economics* 131(1):30–43. doi:10.1016/j.ijpe.2010.12.022.

Lundmark, P. n.d. Control room ergonomics with the operator in focus for an attractive collaborative environment. Zurich: ABB.

Lynas, D. and T. Horberry. 2011. A review of Australian human factors research and stakeholder opinions regarding mines of the future. *Ergonomics Australia – HFESA Conference Edition 2011 11:13.*

Lynas, D. and T. Horberry. 2011. Human factor issues with automated mining equipment. *Ergonomics Open Journal* 4:74–80. doi:10.2174/1875934301104010074.

Maier, M. S., T. Kuhlmann and J. C. Thiele. 2014. Adopting lean and characteristic line based industrial methods for optimizing room and pillar processes. Paper presented at: *Aachen International Mining Symposia*, 11–12 June Aachen.

Mårtensson, L. 1995. Requirements on work organisation: From work environment design of *"Steelworks 80"* to human-machine analysis of the aircraft accident at Gottröra, *Ph.D. Dissertation*, Stockholm, Royal Institute of Technology.

Maslow, A.H. 1987. *Motivation and Personality*. New York: Harper & Row.

Mayo, E. 1945. *The Social Problems of an Industrial Civilization*. Boston, MA: Harvard University.

McBride, D. I. 2004. Noise-induced hearing loss and hearing conservation in mining. *Occupational Medicine* 54(5):290–296. doi:10.1093/occmed/kqh075.

McPhee, B. 2004. Ergonomics in mining. *Occupational Medicine* 54(5):297–303. doi:10.1093/occmed/kqh071.

Mellström, U. 2004. Machines and masculine subjectivity, technology as an integral part of men's life experiences. *Masculinities and Technology* 6(4):368–383. doi:10.1177/1097184X03260960.

Metall. 1985. Det goda arbetet (The good work). Main report from the programme committee on the value and conditions of industrial work presented at the Swedish Metalworkers' Union conference, 1–7 September 1985.

Mottola, L., M. Scoble and M. G. Lipsett. 2011. Machine monitoring and automation as enablers of lean mining. In: S. Saydam ed., *Second International Future Mining Conference*, Sydney, Australia, 22–23 November 2011, The Australasian Institute of Mining and Metallurgy, Victoria, pp. 81–86.

Muther, R. and J. D. Wheeler. 1994. *Simplified Systematic Layout Planning*. Kansas City, MO: Management and Industrial Research Publications.

Muzaffar, S., K. Cummings, G. Hobbs, P. Allison and K. Kreiss. 2013. Factors associated with fatal mining injuries among contractors and operators. *Journal of Occupational and Environmental Medicine* 55(11):1337–1344. doi:10.1097/JOM.0b013e3182a2a5a2.

Nachreiner, F., P. Nickel and I. Meyer. 2006. Human factors in process control systems: The design of human–machine interfaces. *Safety and Design* 44(1):5–26. doi:10.1016/j.ssci.2005.09.003.

Nahrgang, J. D., F. P. Morgeson and D. A. Hofmann. 2010. Safety at work: A meta-analytic investigation of the link between job demands, job resources, burnout, engagement, and safety outcomes. *Journal of Applied Psychology* 96(1):71–94. doi:10.1037/A0021484.

Nakajima, S. 1988. *TPM–An Introduction to Total Productive Maintenance*. Cambridge, MA: Productivity Press.

National Board of Occupational Safety and Health. 1983. Occupational Injuries 1980. Official statistics of Sweden. Stockholm: Statistics Sweden.

National Board of Occupational Safety and Health. 1984. Occupational Injuries 1981. Official statistics of Sweden. Stockholm: Statistics Sweden.

National Board of Occupational Safety and Health. 1985. Occupational Injuries 1982. Official statistics of Sweden. Stockholm: Statistics Sweden.

National Board of Occupational Safety and Health. 1986. Occupational Injuries 1983. Official statistics of Sweden. Stockholm: Statistics Sweden.

National Board of Occupational Safety and Health. 1987. Occupational Injuries 1984. Official statistics of Sweden. Stockholm: Statistics Sweden.

Nelson, M. G. 2011. Evaluation of mining methods and systems. In *SME Mining Engineering Handbook*, ed. P. Darling, vol. 1, 3rd ed., 341–348. Englewood, CO: Society for Mining, Metallurgy, and Exploration.

Nelson, M. G. 2011. Evaluation of mining methods and systems. In *SME Mining Engineering Handbook*, ed. P. Darling vol. 1, 3rd ed.: 341–348. Englewood, CO: Society for Mining, Metallurgy, and Exploration.

Nelson, M. G. 2011. Mine economics, management, and law. In *SME Mining Engineering Handbook*, ed. P. Darling, vol. 1, 3rd ed.: 297–307. Englewood, CO: Society for Mining, Metallurgy, and Exploration.

Neumann, P. 2006. *Inventory of Tools for Ergonomic Evaluation*. Stockholm: Arbetslivsinstitutet.

Neumann, W. P. and J. Dul. 2010. Human factors: Spanning the gap between OM and HRM. *International Journal of Operations & Production Management* 30(9):923–950. doi:10.1108/01443571011075056.

Newsome, K. 2003. The women can be moved to fill in the gaps: New production concepts, gender and suppliers. *Gender, Work & Organization* 10(3):320–341. doi: 10.1111/1468-0432.00198.

NIOSH. 2011a. Metal operator mining facts-2008. NIOSH Publication No. 20111-164. Pittsburgh, PA: National Institute for Occupational Safety and Health.

NIOSH. 2011b. Mining facts-2008. NIOSH Publication No. 2011-161. Pittsburgh, PA: National Institute for Occupational Safety and Health.

Noort, D. and P. McCarthy. 2008. The critical path to automated underground mining. Paper presented at *The First Future Mining Conference*, 19–21 November, Sydney.

Nygren, M. 2016. Coordinating occupational health and safety: Regulatory demands and practical implementation on multi-employer worksites. Paper presented at: 8th *Nordic Working Life Conference* 2016, 2–4 November, Tampere, Finland.

Nygren, M., M. Jakobsson, E. Andersson and B. Johansson. 2017. Safety and multi-employer worksites in high-risk industries: An overview. *Relations Industrielles/Industrial Relations* 72(2):223–245. doi:10.7202/1040399ar.

Ohno, T. 1988. *Toyota Production System: Beyond Large-Scale Production*. Portland, OR: Productivity Press.

Oldroy, G. C. 2015. Meeting mineral resources and mine development challenges. Paper presented at: *Aachen Fifth International Mining Symposia: Mineral Resources and Mine Development*, 27–28 May, Aachen.

Olofsson, J. 2010. Taking place–Augmenting space, *Ph.D. Dissertation*, Luleå University of Technology.

Patterson, J. M. and S. A. Shappell. 2010. Operator error and system deficiencies: Analysis of 508 mining incidents and accidents from Queensland, Australia using HFACS. *Accident Analysis and Prevention* 42(4):1379–85. doi:10.1016/j.aap.2010.02.018.

Peterson, R. A. 1965. Alienation and freedom: The factory worker and his industry, book review. *The Sociological Quarterly* 6(1):83–85.

PricewaterhouseCoopers. 2012. *Mind the Gap: Solving the Skills Shortages in Resources*. n.p.: PricewaterhouseCoopers.

Quinlan, M. 2014. *Ten Pathways to Death and Disaster: Learning from Fatal Incidents in Mines and Other High Hazard Workplaces*. Sydney: Federation Press.

Radjiyev, A., H. Qiu, S. Xiong and K. H. Nam. 2015. Ergonomics and sustainable development in the past two decades (1992–2011): Research trends and how ergonomics can contribute to sustainable development. *Applied Ergonomics* 46:67–75. doi:10.1016/j.apergo.2014.07.006.

Radnor, Z. J. and R. Boaden. 2004. Developing an understanding of corporate anorexia. *International Journal of Operations & Production Management* 24(4):424–440. doi:10.1108/01443570410524677.

Ranängen, H. and Å. Lindman. 2017. A path towards sustainability for the Nordic mining industry. *Journal of Cleaner Production* 151:43–52. doi:10.1016/j.jclepro.2017.03.047.

Randolph, M. 2011. Current trends in mining. In *SME Mining Engineering Handbook*, ed. P. Darling, vol. 1, 3rd ed.: 11–19. Englewood, CO: Society for Mining, Metallurgy, and Exploration.

Randolph, M. 2011. Current trends in mining. In: *SME Mining Engineering Handbook*, ed. P. Darling, vol. 1, 3rd ed.: 11–19. Englewood, CO: Society for Mining, Metallurgy, and Exploration.

Ranhagen, U. 1995. *Människa, miljö, mål: utvecklande arbete, säkrare arbetsplatser, effektivare företag (Human, Environment, Goals: Developing Work, Safer Workplaces, More Effective Businesses)*, vol. 6, *Att arbeta i projekt: om planering och projektering inför större förändringar* (To work in projects: on planning and projects for larger changes). Stockholm: Arbetarskyddsnämnden.

Ranhagen, U. 2004. Planering för en god arbetsmiljö och ett hållbart arbetsliv (Planning for a good work environment and a sustainable working life). In *Framtidens arbetsmiljö- och tillsynsarbete (Work environmental and supervisory work of the future)*, eds. B. Johansson, K. Frick, and J. Johansson, 284–303. Lund: Studentlitteratur.

Rasmussen, B. 1999. Dehierarchization–Re-organizing gender? *Ph.D. Dissertation*, Norges teknisk-naturvitenskaplige universitet.

Reason, J. 1997. *Managing the Risks of Organizational Accidents*. Surrey: Ashgate.

Reeves, E. R., R. F. Randolph, D. S. Yantek and J. S. Peterson. 2009. *Noise Control in Underground Metal Mining*. Pittsburgh: The National Institute for Occupational Safety and Health. www.cdc.gov/niosh/mining/UserFiles/works/pdfs/2010-111.pdf.

Romero, D., J. Stahre, T. Wuest, O. Noran, P. Bernus, Å. Fast-Berglund and D. Gorecky. 2016. Towards an operator 4.0 typology: A human-centric perspective on the fourth industrial revolution technologies. In *Proceedings International Conference on Computers & Industrial Engineering (CIE46)*.

Røvik, K. A. 2000. *Moderna organizationer: Trender i organizationstänkandet vid millennieskiftet (Modern Organisations: Trends in Organisational Though at the Millennium Shift)*. Malmö: Liber.

Ruff, T., P. Coleman and L. Martini. 2011. Machine-related injuries in the US mining industry and priorities for safety research. *International Journal of Injury Control and Safety Promotion* 18(1):11–20. doi:10.1080/17457300.2010.487154.

Ruiz Martín, A., M. Rodríguez Díaz and J. A. Ruíz San Román. 2014. Measure of the mining image. *Resources Policy* 41(1):23–30. doi:10.1016/j.resourpol.2014.01.004.

Safe Work Australia. 2011. *How to Manage Work Health and Safety Risks: Code of Practice*. Gosford, NSW: WorkCover NSW.

Saleh, J. H. and A. M. Cummings. 2011. Safety in the mining industry and the unfinished legacy of mining accidents: Safety levers and defense-in-depth for addressing mining hazards. *Safety Science* 49(6):764–77. doi:10.1016/j.ssci.2011.02.017.

Saurin, T. and C. Ferreira. 2009. The impacts of lean production on working conditions: A case study of a harvester assembly line in Brazil. *International Journal of Industrial Ergonomics* 39(2):403–412. doi:10.1016/j.ergon.2008.08.003.

Schmitz Jr., J. A. 2005. What determines productivity? Lessons from the dramatic recovery of the U.S. and Canadian iron ore industries following their early 1980s crisis. *Journal of Political Economy* 113(3):582–625. doi:10.1086/429279.

Schonberger, R. 1982. *Japanese Manufacturing Techniques: Nine Hidden Lessons in Simplicity*. New York: Free Press.

Shingo, S. 1981. *A Study of the Toyota Production System*. New York: Productivity Press.

Shooks, M., B. Johansson, E. Andersson and J. Lööw. 2014. *Safety and Health in European Mining.* Luleå: Luleå University of Technology.

Simpson, G., T. Horberry and J. Joy. 2009. *Understanding Human Error in Mine Safety.* Surrey: Ashgate.

Smith, M. J. and P. Carayon-Sainfort. 1989. A balance theory of job design for stress reduction. *International Journal of Industrial Ergonomics* 4(1):67–79. doi:10.1016/0169-8141(89)90051-6

Somerville, M. and L. Abrahamsson. 2003. Trainers and learners constructing a community of practice. *Studies in the Education of Adults* 35(1):19–34. doi:10.1080/02660830.20 03.11661472.

Statistics Sweden. 1982. Statistical abstract of Sweden 1982/83. Official Statistics of Sweden. Stockholm: Statistics Sweden.

Statistics Sweden. 1983. Statistical abstract of Sweden 1984. Official Statistics of Sweden. Stockholm: Statistics Sweden.

Statistics Sweden. 1984. Statistical abstract of Sweden 1985. Official Statistics of Sweden. Stockholm: Statistics Sweden.

Statistics Sweden. 1985. Statistical abstract of Sweden 1986. Official Statistics of Sweden. Stockholm: Statistics Sweden.

Statistics Sweden. 1986. Statistical abstract of Sweden 1987. Official Statistics of Sweden. Stockholm: Statistics Sweden.

Statistics Sweden. 1996. *Bakgrundsmaterial om befolkningens utbildning: Utbildningsnivå i kommunerna, länen och riket 1995-01-01 och 1996-01-01 (Background material regarding the education of the population: Educational levels in the municipalities, counties and nation 1995-01-01 and 1996-01-01).* Stockholm: Statistics Sweden.

Statistics Sweden. 2016a. *Labour Statistics Based on Administrative Sources.* Örebro: Statistics Sweden. www.scb.se/am0207-en.

Statistics Sweden. 2016b. *The Swedish Occupational Register with Statistics.* Örebro: Statistics Sweden. www.scb.se/am0208-en.

Statistics Sweden. n.d. Medarbetarundersökningar (Employee surveys). (webpage). Available at: www.scb.se/vara-tjanster/insamling-och-undersokning/medarbetarundersokningar/ (Accessed 2018-05-07).

Steinberg, J. G. and G. De Tomi, G. 2010. Lean mining: Principles for modelling and improving processes of mineral value chains. *International Journal of Logistics Systems and Management* 6(3):279–298. doi:10.1504/IJLSM.2010.031982.

SveMin. 2010. *Occupational Injuries and Sick Leave in the Swedish Mining and Mineral Industry 2009.* Stockholm: SveMin.

SveMin. 2016a. *Bränder och brandtillbud 2016 (Fires and Fire Incidents 2016).* Stockholm: SveMin.

SveMin. 2016b. *Occupational Injuries and Sick Leave in the Swedish Mining and Mineral Industry 2015.* Stockholm: Sweden.

Swedish Work Environment Authority. 2011. Occupational accidents and work-related diseases 2010. Arbetsmiljöstatistik Rapport 2011:1. Stockholm: Swedish Work Environment Authority.

Swedish Work Environment Authority. 2012. Occupational accidents and work-related diseases 2011. Arbetsmiljöstatistik Rapport 2012:2. Stockholm: Swedish Work Environment Authority.

Swedish Work Environment Authority. 2013. Occupational accidents and work-related diseases 2012. Arbetsmiljöstatistik Rapport 2013:1. Stockholm: Swedish Work Environment Authority.

Swedish Work Environment Authority. 2014. Occupational accidents and work-related diseases 2013. Arbetsmiljöstatistik Rapport 2014:1. Stockholm: Swedish Work Environment Authority.

Swedish Work Environment Authority. 2015. Arbetsskador 2014 (Occupational accidents and work-related diseases 2014). Arbetsmiljöstatistik Rapport 2015:1. Stockholm: Swedish Work Environment Authority.

Swedish Work Environment Authority. 2016. Occupational accidents and work-related diseases 2015. Arbetsmiljöstatistik Rapport 2016:1. Stockholm: Swedish Work Environment Authority.

Swedish Work Environment Authority. 2017. Occupational accidents and work-related diseases 2016. Arbetsmiljöstatistik Rapport 2017:1. Stockholm: Swedish Work Environment Authority.

Taylor, B. W. K. 2006. A feminist critique of Japanization: Employment and work in consumer electronics. *Gender, Work & Organization* 13(4):317–337. doi:10.1111/j.1468-0432.2006.00310.x.

Taylor, F. W. 1911. *The Principles of Scientific Management*. New York: Harper & Brother.

Thompson, P. and C. Warhurst, eds. 1998 *Workplaces of the Future*. London: Macmillan Press.

Thorsrud, E. and F. Emery. 1969. *Mot en ny bedriftsorganisation (Towards a new company organization)*. Oslo: Tanum forlag.

Tilton, J. E. 2010. Is mineral depletion a threat to sustainable mining? *SEG Newsletter* 82. Littleton, CO: Society of Economic Geologists.

Tilton, J. E. 2014. Cyclical and secular determinants of productivity in the copper, aluminum, iron ore, and coal industries. *Mineral Economics* 27(1):1–19. doi:10.1007/s13563-014-0045-9

Trist, E. L. and K. W. Bamforth. 1951. Some social and psychological consequences of the longwall method of coal-getting: an examination of the psychological situation and defences of a work group in relation to the social structure and technological content of the work system. *Human Relations* 4(1):3–38. doi:10.1177/001872675100400101.

Trist, E. L., G. I. Susman and G. R. Brown. 1977. An experiment in autonomous working in an American underground coal mine. *Human Relations* 30(3):201–36. doi:10.1177/001872677703000301.

Ventyx. 2012. *Ventyx Mining Executive Insights Annual Global Survey Results 2012 (White Paper)*. n.p.: Ventyx.

Vink, P., E. A. Koningsveld and J. F. Molenbroek. 2006. Positive outcomes of participatory ergonomics in terms of greater comfort and higher productivity. *Applied Ergonomics* 37(4):537–546. doi:10.1016/j.apergo.2006.04.012.

Wärvik, G. and P. O. Thång. 2003. Conditions for learning from perspective of labor workers in connection with "the new work order". Paper presented at: *3rd International Conference of Researching Work and Learning*, 25–27 July, Tampere, Finland.

Wenger, E. 1998. *Communities of Practice*. Cambridge, NY: Cambridge University Press.

Westgaard, R. H. and J. Winkel. 2011. Occupational musculoskeletal and mental health: significance of rationalization and opportunities to create sustainable production systems–A systematic review. *Applied Ergonomics* 42(2):261–96. doi:10.1016/j.apergo.2010.07.002.

Wibowo, A. P. and F. A. Rosyid. 2008. Development concept in the mining industry – An accelerator to achieve regional sustainable development. Paper presented at: First International Future Mining Conference, 19–21 November, Sydney.

Wicks, D. 2002. Institutional base of identity construction and reproduction: The case of underground coal mining. *Gender, Work and Organization* 9(3):308–335. doi:10.1111/1468-0432.00162.

Widzyk-Capehart, E. and E. Duff. 2007. Automation: A blessing or a curse? Paper presented at *Sixteenth International Symposium on Mine Planning and Equipment Selection*, 11–13 December, Bangkok.

Wijaya, A. R., R. Kumar and U. Kumar. 2009. Implementing lean principle into mining industry - issues and challenges. In: R. K. Singhal ed., *18th International Symposium on Mine Planning and Equipment Selection*, Banff, Canada, 16–19 November 2009, Reading Matrix Inc.

Wikberg-Nilsson, Å. and S. Johansson. 2010. Reflection-for-action: A collaborative approach to the design of a future factory. *Proceedings of 3rd International Conference on Applied Human Factors and Ergonomics, AHFE,* 15–18 July 2010, Miami, FL.

Wikberg-Nilsson, Å., Å. Ericson and P. Törlind. 2015. *Design: Process och metod (Design: Process and Method).* Lund: Studentlitteratur.

Wikberg-Nilsson, Å., L. Abrahamsson, Y. Fältholm, B. Johansson, J. Johansson, S. Johansson and K. Rask. 2011. *Framtidsfabriken: En vision av framtidens effektiva och attraktiva arbetsmiljöer i industrin (The Future Factory: A Vision of the Future's Effective and Attractive Industrial Work Environment).* Luleå: Luleå University of Technology.

Wikberg-Nilsson, Å., Y. Fältholm and L. Abrahamsson. 2009. Effective use of human actors in change process: A participative approach to industrial development. *Proceedings of SPS 09,* December 2009, Gothenburg, Sweden.

Womack, J. P., D. T. Jones and D. Roos. 1991. *The Machine That Changed the World.* New York: Macmillan Publishing Company.

Womack, J.P. and D. T. Jones. 2003. *Lean Thinking.* New York: Macmillan Free Press.

Yingling, J. C., R. B. Detty and J. J. Sottile. 2000. Lean manufacturing principles and their applicability to the mining industry. *Mineral Resources Engineering* 9(2):215–238.

Zacharatos, A., J. Barling and R. D. Iverson. 2005. High-performance work systems and occupational safety. *The Journal of Applied Psychology* 90(1):77–93. doi:10.1037/0021-9010.90.1.77.

Zappala, J. 1988. Just-in-time techniques in metal manufacturing: an assessment of its implementation and impact. Occ. Paper No. 5, Dept. of Industrial Relations, University of Sydney, Sydney.

Zhang, M., V. Kecojevic and D. Komljenovic. 2014. Investigation of haul truck-related fatal accidents in surface mining using fault tree analysis. *Safety Science* 65:106–117. doi:10.1016/j.ssci.2014.01.005.

Zhang, T. and M. A. Barclay. 2007. *What Students Want: Career Drivers, Expectations and Perceptions of Mining Engineering and Minerals Processing Students.* Sydney: Minerals Council of Australia.

Index